LOT'S WIFE

LOT'S WIFE

SALT AND THE HUMAN CONDITION

SALLIE TISDALE

HENRY HOLT AND COMPANY

NEW YORK

Published by Henry Holt and Company, Inc.,
115 West 18th Street, New York, New York 10011.
Published in Canada by Fitzhenry & Whiteside Limited,
195 Allstate Parkway, Markham, Ontario L3R 4T8.

Library of Congress Cataloging-in-Publication Data
Tisdale, Sallie.
Lot's wife.
Bibliography: p.
1. Salt in the body.   2. Salt—Miscellanea.
I. Title.
QP535.N2T57   1988   612'.01524   88-8798
ISBN 0-8050-0920-5

Henry Holt books are available at special discounts
for bulk purchases for sales promotions, premiums,
fund raising, or educational use. Special editions
or book excerpts can also be created to specification.

For details, contact:

Special Sales Director
Henry Holt and Company, Inc.
115 West 18th Street
New York, New York 10011

First Edition

Designer: Ann Gold
Printed in the United States of America

1   3   5   7   9   10   8   6   4   2

Grateful acknowledgment is given for permission to use excerpts from the
following poems:
   "Sea Fever" and "Roadways," by John Masefield, from *The Poems and
Plays of John Masefield* by John Masefield (New York: Macmillan, 1953),
reprinted with permission of Macmillan Publishing Company.
   "Ode to Salt," by Pablo Neruda, from *News of the Universe*, edited by
Robert Bly, copyright © 1980 by Robert Bly, reprinted with permission of
Sierra Club Books.

For Judy,
kind hands and whole heart

ACKNOWLEDGMENTS

Once again, I've been blessed with the voice of an excellent editor, Channa Taub, who holds this work close to her own heart and deftly parries the concern of people who wonder what in the world her writers are doing. Hers are the words between the lines.

A number of people listened, sometimes patiently, to my unformed ideas and loose phrases. Chief among them are Katinka Matson, Lenny Dee, and my husband, Bob Macer, who has an ear for a good story. Sy Safransky and Kevin Kelly chipped in with important points at the last minute. Father Edmond Bliven, Rabbi Yonah Geller, and Dr. Bill Randall gave expert advice.

The "old men of salt," not all of them old, initiated me into a few of the mysteries. They include Jim Palmer, Greg Bottelberghe, Roger Peterson, Dr. Wallace Gwynn, and Peter Czerny.

Thanks to the people who, upon discovering I was writing a book about salt, said, "Have I got a story for you."

For Judy,
kind hands and whole heart

ACKNOWLEDGMENTS

Once again, I've been blessed with the voice of an excellent editor, Channa Taub, who holds this work close to her own heart and deftly parries the concern of people who wonder what in the world her writers are doing. Hers are the words between the lines.

A number of people listened, sometimes patiently, to my unformed ideas and loose phrases. Chief among them are Katinka Matson, Lenny Dee, and my husband, Bob Macer, who has an ear for a good story. Sy Safransky and Kevin Kelly chipped in with important points at the last minute. Father Edmond Bliven, Rabbi Yonah Geller, and Dr. Bill Randall gave expert advice.

The "old men of salt," not all of them old, initiated me into a few of the mysteries. They include Jim Palmer, Greg Bottelberghe, Roger Peterson, Dr. Wallace Gwynn, and Peter Czerny.

Thanks to the people who, upon discovering I was writing a book about salt, said, "Have I got a story for you."

CONTENTS

May my harvest fail,
my cattle die,
and may I never taste salt again,
if I do not speak the truth.

—Sumatran saying

PART I
UNIVERSAL

1
OCEAN

THERE IS NO salt in the ocean. Dash, splash, wave on wave and the sportive fingers of spindrift between your toes: no salt. Salt dissolves in water, you know, the crystals unraveling into ions of sodium and chloride floating free. The spray in your face is a soup of potential, electric with promise; the fine and secret process of evaporation vibrates with anticipation behind the veil. Roar, the wave breaks, curls in briny foam, the sweet fishy froth wets your mouth. The water skims away into the air, and the little skittery atoms coalesce. And then salt emerges and follows you home, like sand in your shoes.

What we call salt is sodium chloride, the steadfast marriage of two otherwise inconstant creatures: the base metal sodium and the acidic gas chlorine. Any acid and any base in tandem make a salt, and there are many such salts split in the ocean. Salt melts water, cleaves in silence, the ballet of electron shells transforming one molecule into another. We call sodium chloride salt, and the crystal face of sodium chloride—bright, white, light—floats like a ghost through the ocean. We can't imagine one without remembering the other.

Some of us like the shore and the shallows and some prefer the deeps, the unrepentant mystery disappearing beneath dangled toes and an angled rudder: ocean without apology, uncompromised and deadly, salty, scary, cool. Boundless and unstilled ocean, warm and tickling tides—and tide means time, a fixed flood and ebbing flow. The sea surface mirrored blue on a silent day is suddenly rippled by a cat's-paw wind. The glassy water bubbles with a rapid passage of tiny cresting waves that drop suddenly away to flatness again: the precursor of a storm. And always and in everything the sandy grit of salt: drying in your hair till it hangs in wind-made braids like ocean dreadlocks, powdering your cheeks and roughing your knuckles. It chafes your chin till the skin glows pink with a salt glow, old remedy for sloth. There is no salt in the ocean, but salt is there, in the depth and in the barely visible spray that makes rainbows of the wave crests and distorts the curve of the earth's horizon. Salt is the blood of the ocean, the bath of the ocean's cells, electric sparkle of the sea's heartbeat, stimulus to the sea's contracting muscle. Swim in the sea, go on: your skin is very thin. Blood meets blood. You're right at home.

Iceland, greece, China, Norway, Japan. The story is the same, and very old.

For reasons unclear in the stretch of time but perhaps no more complicated than the generosity of the affluent, one of the small people appeared to a peasant. The peasant's family was hungry and ill-clothed, though his brother was a wealthy merchant who dined on delicacies. The elf brought a gift to the poor man to even the balance, a magic millstone that ground out desires.

"Turn the stone to the right," he instructed the peasant, "and the stone will give you what you want. When you have enough, turn the stone to the left and it will stop."

The peasant was delighted; his hopeless prayers were answered. He turned the stone and ground food, clothes, a house. He ground out gold coins. His children were fed; his wife warm and able to rest a while.

The rich brother, of course, was angered by the turn—so to speak—of events. Where did the sudden riches come from? Did his brother know a secret? And so he spied upon him, and saw him turn the stone late one night for a vessel of wine, and vowed to steal it.

Soon he found his chance, because the good peasant lacked the suspiciousness of the wealthy. The brother rolled the stone from the house, down the hill to the bay, and into a boat. He rowed out to sea to run away with his treasure, unafraid. It was night, the ocean was deep and dark, but with a magic millstone he felt safe enough.

He grew hungry, and turned the stone to the right, commanding it to make food. The stone turned, food appeared. But tasting it the brother was (of course!) dissatisfied. "It needs salt," he grumbled to the stone. "Make me salt." And the stone obeyed.

Oh, but the brother had grabbed the stone too soon. He hadn't waited to see how to make the stone stop. "Enough!" he shouted when a pile of salt lay at his feet. "Stop!" But the stone didn't stop. Instead it turned and turned, steady and sure, salt spilling from its center like snow. And in the middle of the night in the middle of the ocean the boat filled with salt and sank, and the brother drowned. The stone sank too, spinning down. It rolls along the seafloor still, pouring out salt into the ocean, into the salty salty waters of the sea.

THE OCEAN has always been a puzzle, a mystery, impenetrable. The ancients used to sit at the shore and ponder not only the ocean's origin, but its purpose, its meaning, how it came to earth and how it came to be salt. Some thought the ocean a product of underground springs, bubbling through salty rocks to the surface from some unknown, eternal source. Others thought it the condensed product of an envelope of moisture that once surrounded the earth like an eggshell, an infinite flowing stream called Okeanos. A vocal minority proclaimed it sweat, exuded by the earth in the sun's raw heat, and claimed that accounted for its saltiness, too: Isn't all sweat salty? The more religious among the philosophers and the less academic among the people had lustier ideas: The ocean isn't the earth's sweat, but God's—or God's urine. (And in turn the sun and moon are spangles of urine splashed up from the spray as it struck the dirt.) Others thought it milk, a nurturing discharge from earth's nipple. And there is always Genesis: With one stroke Yahweh sliced the water loose, and let it flow.

Aristotle, as usual, had a few things to add. He takes each notion in his hand the way he took jawbones and eyeballs, and dissects it: "Metaphors are poetical and so that expression may satisfy the requirements of a poem," he wrote about the idea of ocean as sweat. "But as a scientific theory it is unsatisfactory."

Aristotle called the ocean bowl the natural place of water— water's end, and not its source: "The finest and sweetest water is every day carried up and is dissolved into vapour and rises to the upper region, where it is condensed again by the cold and so returns to the earth." The sun's heat leaves salt behind in the ocean like the dregs in a cup of mead, and the salt itself is just the earth's "undigested residue," the stuff that makes ocean water heavier than fresh, thicker, more buoyant. Salt is the ocean's "savor," and its heat.

All quite correct. All of it: residue and urine, sweat and spangling planets shining overhead, savor and eggshells of humidity above. The wave rolls in and under my toes the sand comes alive, curling from underneath my feet like a school of fish caught out. The sun's glare on the white crests makes me blink in time to catch the broken wave full in the face. I sputter and sing. Where did you come from, ocean, and where your salt?

THE OCEAN began fresh, claim the new ancients, philosophers called scientists. This is the province of the geologist, guesswork of ages. The ocean began as the steady condensation of a wet and violent atmosphere, roiling with volcanic dirt and torn by cataclysmic storms that slowly stilled, drop by drop.

The water vapor itself evolved first, released in chemical pieces from rocks and hot springs and vents. (Such a thing is called juvenile water, a nod, perhaps, to its naive leap into the universe.) Water, after all, is only one way to combine the simple elements of hydrogen and oxygen, a product of temperature and pressure and periodicity—a serendipity of the forces of time. The steady rain that dribbled into the rough basins washed chlorine gas from the smoke, and grabbed young minerals, sodium and calcium and potassium, from igneous rock, washed them down the riverbeds to the sea. So water came, and minerals and gas, and together they wet the low places of a jagged planet, salting it like broth.

A fine broth, too: after a spell our ancestors crawled out, shook the water from their scales, and took up residence. My blood is like seawater—but not the seawater I play in today. As a child I imagined the watery course of the unborn, growing and then losing a tail and gills. I could see my skin as a sheet of refined scales, and my toes a logical step up from flippers. Now I find my blood is a Precambrian composition, minerals in balance like the seawater of millennia ago, revealed in the sediment and fossil of rational guesswork. Taken as a sack of minerals, I reflect my own evolution, a geological period winked in the universe's eye. In 1926 a man named Macallum invented a word for the idea: *paleoratio*, the ontological nod to the phylogene.

"And the flung spray and the brown spume, and the seagulls crying." That's the chantey, the siren call of sailors to the sea, the sea. It is more than two-thirds of our world, 18 times greater in volume than the wee bit of land we huddle on. The sea rocks our cradle of earth like a bobbing cork, pushing continents closer, mountains up, islands down. Big: 330 billion cubic miles of water, each cubic mile with 4 billion tons of oxygen, 166 million tons of salt, and varying weights of several dozen rare and precious things: argon and cerium, titanium, krypton, silver and gold.

There is enough salt in the ocean to cover the continents

500 feet thick. Great beds of salt like blocks of fertilizer—plankton food, ocean blood—are buried in the ocean floor. And still water and minerals are being poured into the ocean at a frightful rate, barely making a difference. By all appearances the ocean's composition hasn't changed in a few eons or so: but what a fractional bit of time we've spent checking.

The ocean is rich, but not full—there's plenty of room for more. There are limiting factors at work; the making of salts at water's edge, minerals adsorbing to clays and forming new compounds, sinking to the bottom and pressing into the bed, chewed up by plankton, fish, and plants. All in all the salinity of the ocean—a measure of sodium and chloride and all the dissolved compounds—is about 3.5 percent. It varies, but round off the extremes and there are 35 parts salt to 1,000 parts water. Salinity depends on temperature, pressure, density, the peculiarities of geology in the crust and shore—and still the ocean's salt stays about the same. A little more here and a little less a few thousand miles away, but the ocean taken as a whole—and how else can it be taken?—seems to have a correct equation for itself, a proper proportion that asserts itself as certainly as the tides. Swimming in the oceans is a vast amount of that electric, eclectic sodium: 14.1×10^{15} tons of it—that's 14,100,000,000,000,000 tons. Each year another 205,000,000 tumble in, from river and rain, and each year about that much tumbles out, crawling up into rock like the first crawling amphibians, sinking to the fertile bricks of salt below, splashed in your hair, clinging to my sleeve. Sodium gets around, and a few people have tried to follow it on its trek.

Geologists in the past assumed the ocean had to grow saltier with time, with every season; they shared Aristotle's concept of corruption. How could salt rise and fall, when evaporation pulls out the fresh? And if the sea is getting saltier, then couldn't a few measurements help us trace back to its origin, by figuring its sweet beginning? Richard Halley, of comet fame, proposed measuring the sodium content of the sea every ten years, and figuring backward. His imagination failed him; ten years is such a short, short time that even a large difference could not be discerned. He was distracted, anyway, and the project wasn't undertaken until 1899, by an Irishman named John Joly.

Joly guessed the ocean's age, based on the sodium content, at 100 million years—far too few. He found it impossible to embrace the real cycle, the dizzying spin of centuries: sodium's return and exchange, just like the come and go of oxygen, proteins, and the water molecules themselves. In a true chemical cycle, any single molecule may travel for a million years, from water to rock to mountain, eroded and washed away, blown across plains in the wind, buried and evaporated and eaten, processed out and leached back, dissolved and sunk and borne away into a compound, a cell wall, a bone, excreted in a shark's urine—again and again. Joly, it turns out in the current paradigm, came up with a figure that was about half the time a single ion of sodium spends in going from earth to ocean and back.

For a long while after Aristotle the ocean was thought of as a bowl—a vessel of sloping sides, receiving. Is it such an odd notion to imagine it growing saltier with time? (And larger, perhaps, in the wake of floods and changing coastlines that wiped away villages, cultures, races?) The very idea of atoms was hard. Even in the 1930s scientists couldn't venture past 200 million years, quaking at the unimaginably small size of an ion, these astonishing volumes, this extraordinary period of time.

It wasn't until 1960 that a Norwegian named Tom Barth spoke out for cycles, instead of gain. The ocean to Barth was a tube instead of a bowl—a channel. He was a geologist undeterred by the intimidating ages. Barth figured, through a tangled complexity of mathematics and geological statistics, that it took about 208 million years for a single piece of sodium (tiny thing, that) to find its way round, and furthermore it would take at least three such cycles to bring the ocean to the kind of stability it seems to be enjoying now: three times 208 million years of evolutionary turmoil to let the ocean's salt settle out, settle down, stay put. Sodium cycles, plate tectonics, new notions of magnetism: all have changed the vision of the earth, from a solid basin wet in its depths to a shifting mass of eternal atoms in constant change, each piece reworking itself in and around and in and upon itself again and again and again. How old the ocean? How old its salt as fresh and sharp as a hand slap, as new as dawn? Six hundred million years. Two billion years, three billion years. Hop in, there's room, and lots of time. Do you know the riddle? "I threw it in the water, it changed into water. It lives in the water, yet water kills it." What could it be? What but salt?

NEAR MELTING ice, ocean water is fresher; near freezing ice it grows saltier, for water freezes fresh. The water left behind gets thicker, denser, and its freezing point drops. A little more freezes out and the rest is saltier still, and colder before it grows solid. Cold water contracts, and grows even denser, and after a while begins to sink, cold and salty currents falling below the Antarctic shelf like a heavy syrup falling through a lighter one. When this bundle hits bottom it begins to crawl, an internal wave hundreds of feet high, layered cold and salty in a blanket of warmer, sweeter water. The wave slithers north along the bottom, very slow, getting older, staying cold and salty till it reaches the tropics. Up above are complex currents and tides born from temperature, bottom slope, and wind direction, radiating into each other like waves from thrown pebbles, complicated things that may pass south to north and back in a few years' time. On the bottom several hundred years may pass in the journey before the ocean grog reaches the equator, growing lighter, fresher, and starts to rise. The wave you surf through now has been buried a long, long time—listen to it, that wave's been around. Slow massive furrows swell out, propelling ships and plankton, driftwood, rolling nodules of manganese and whales' earbones, sunken shells and unanchored kelp, rolling them along. Currents are the earth's spin, sun's heat, moon's pull. I'm a soft sack of ocean, blood's blood, with no escape; like you I have my tides, a fraction of an ounce of weight gained and lost, a secret tug south, a restless hunger for the east now, the west tomorrow. My salt cycles too, through sea and rock and albatross.

Oh, the seas, the seas: the Red Sea and the Coral, Euboean and Tyrrhenian and Bali, the Ross Sea and the Laccadive, Wadden Sea and Azov Sea and Baltic, the Sea of Yellow, Sea of White, the Black Sea and the Aral, the Arafura, Arctic, Okhokst and Irish, Sulu and Indian and Pacific and more. Float in the Dead Sea buoyant as a cork, bobbing like wood or jellyfish, bobbing as though the sea were a drug. And it is: all a bit dizzyingly large, outside the framework. But sanctuary still, taken skin to skin, hand to mouth: "leads me, lures me, calls me to salt green tossing sea, a mad salt sea-wind blowing the salt spray in my eyes." Sailor songs, chanteys of rhythm, and work, and love. Odes to the fundamental, to this great uterine surf, gravy and ghee, a succulent hydrocele, hemal and moist: the sea baptismal and bloody, earth's womb.

2
FETUS

BIG SEA, LITTLE sea, and all kinds of internal waves. From its first indelible moments the fetus lives with salt, its blood nearly as salty as the blood of an adult, bathing in the salty milk of amniotic fluid. "Salt in the cradle keeps the baby safe," goes the saying, and so the fluid does. The fetus, by its nature a creature of change, rocks in its wet cushion, in its bath of amnion. It swings in the gentle tides of bloodlike ocean water cousin to urine, the sticky yolk of fluid, the way a jellyfish swings to the beat of the current, or an anemone gestures to a passing shrimp: at home.

At the age of two months the fetus is still barely a half-inch long. The legs are hardly there, a mere impression; the length is measured from the crown of the skull to the rump. This runt hiccups, turns its cumbersome, germinating head from side to side. The whole body shimmers reflexively in snakelike waves from head to the suggested tail. It shivers with the unspoken pleasures of pure reciprocity.

The fetus is nothing if not mutual. In these early months of fragility and for months to come, the fetus is a membrane. It not only swims in a growing bubble of fluid, but transfers the fluid through its skin, which lies between the fetal organs and the fluid like a laminary veil. Through filmy eyelids and furred back, the curtain of cells rustles with the busy commuting of molecules. While the skeleton gradually asserts itself, while the ear curls up out of the skull and the fingers slowly flex and expand, the amniotic fluid slides through the fetal skin in a tiny tide of osmosis and pressure, in wavelets of precise exchange.

Amniotic fluid is called liquor. It is composed of water, mostly, like so many things, and more: calcium, magnesium, iron, copper, zinc, more than 20 enzymes, manganese and bilirubin, several hormones, urea and uric acid, proteins, creatinine and glucose, amino acids, potassium, sodium, and chloride. The amount at any time is related to the size and weight of the fetus more than anything else. The volume varies not so much from month to month as day to day—almost hour by hour. If there could be such a thing as an average fetus, that fetus at four months' gestation would be surrounded by nearly a quart of fluid, a pillow, a blanket, a cushion, a cloud.

The fetus grows, its presence is unmistakable. It is slowly distinguishing itself, increasing its discreteness and privacy, drawings its border tighter. If you could touch this thing, it would move in response to your touch, each limb separate. As time passes the skin cells thicken and learn tactile awareness, move from a supposed numbness to sensitivity. Their fairy delicacy gives way to a hard opacity.

Before the fifth month of pregnancy is over the fetal skin is a nearly solid wall. The fetus moves in constant, irregular jerks. It yawns, stretches, nods its head, touches its face. Something begins to awaken, the potential for awareness becomes response. The fetus stirs in the depths like some wide-eyed bottom-dweller sliding, with the currents, toward some lighter place.

From its third month the fetus has been opening its mouth, and waves of amniotic fluid swirl through. It sucks and makes vain attempts to breathe. The tongue and taste organs are already functioning. A few sips of fluid an hour, a few milliliters a day, seem to agree with the fetus; by the end of its term in a bubble of fluid, it drinks with an appetite as voracious as that of the newborn: 450 milliliters of fluid a day, and occasional stray pieces of its own pale hair.

The kidneys are at work, and the fetus urinates. The fetus fills its soft shell with urine and the saltiness suddenly rises, then plateaus. Before birth about 28 milliliters an hour pass into the fluid—about one fluid ounce—all water-clear and slightly salty, a gift of the thing to itself.

The origin and fate of amniotic fluid remains a mystery, in spite of these earnest efforts. Biology is flush with success, with odd and disconcerting facts blank on the page. Sometimes the facts are not enough. But biology's greatest failure is to misunderstand mutuality, in the midst of it. I am glad to know that a fetus turns to root long before it can hope to reach a nipple, but what to make of how I know that? Biological research is like a machining of the spirit.

"The mammalian fetus is well hidden from the inquisitive eye," writes a scientist, "and once exposed, its performance is usually short-lived and not necessarily normal." He bemoans the vagaries of the organism, the lack of cooperation inherent in his work. He is irritable with thwarted curiosity.

Early experiments, and a few more recent ones, describing the movements and reflexes of the fetus have been done on living abortants. This is how we know about its sinusoid shimmering, its reflexive curls. Because a fetus dying on a table from lack of oxygen still turns its cheek to a stroking finger, we know its urge to root is born early, and pure.

One scientist removed fetuses between ten and 22 weeks in gestation, still integral in their amniotic sacs. He aspirated a bit of fluid, to keep it uncontaminated, and then split the sac open over a jar, to capture the wave of salty liquor as it spilled. He took blood from the mother, blood from the fetus, aspirated the fetal urine with a needle in the bladder, and took samples of skin from "the intact fetus." He did this three dozen, five dozen, six dozen times.

Similar research has been done by others, on humans, on dogs and sheep and rabbits and the pouch-breeding kangaroo. I am filled with these pictures, captioned by the stark descriptions. There is such a tone of no-nonsense to it all, a getting-down-to-business that I find more than puzzling. What does it imply, that a man can hold dozens of homunculi in hand, greasy and damp with the womb's smell rising, and reach blithely for the needle?

Abortions are still done with saline water, an injection of salt into the amnion to replace part of the fluid—a salt storm in the fetal universe, sharp and evaporative. Eventually the fetus convulses, labor begins, the unviable is born and dies, if not already dead.

Easy enough to argue necessity. I've argued it myself many times. But I think of all the spilled salt, the old superstitions of salt's primitive powers. It passes through us without conscience or concern and sometimes we dribble it through our fingers like chalk dribbled in a magic design, calling unexpected demons to join our work.

No one knows where amniotic fluid comes from, or where it goes. Something so common, so present, so frequently encountered, arises and disappears with stupefying secrecy. Hippocrates suggested the fluid was a dynamic product of the kidneys, but his common-sense ideas disappeared in the rise of Western science. It was believed, as recently as fifty years ago, that amniotic fluid was a stagnant pool, a puddle in which the drowned, tube-breathing fetus lay. It is clear now that something quite the opposite is true, that the amniotic fluid is a constantly shifting tide, a dynamic volume of tissue fluid flowing to and from the mother and baby by every conceivable route. Fluid is present in the tissues of a freshened embryo when it begins to divide, and is sometimes found in simple ova, unfertilized but ready. (When a fetus dies, as it sometimes does, this continuous transfer of fluid, the tiny tides, slows but doesn't stop. It is a moon's pull of movement.) The exchange appears to be a mechanism unto itself, both independent of the fetus and inextricably linked to it. Fluid recedes from one space, returns silently to top it up again. Some researchers believe the fetus saves a bit of protein from each round of swallowing and urination. What triggers it? A mindless motion of electrical fields, seeking stability, or something less quanti-fiable, less concrete?

This intricate system sometimes fails, as systems will. Sometimes the bubble of fluid swells enormously, filling the uterine vessel till it stretches as if it could burst—a tsunami of the amnion, a flood of fluid. This is called polyhydramnios, and its cause is often obscure. But again and again it is as-sociated with certain failures in the fetus: defects in the mouth and throat that prevent swallowing; malformed kidneys that pour urine out in vast amounts; misshapen skulls that expose the brain, or the lack of one, to the amniotic water. Some babies live, some can't.

Less often there is too little fluid: oligohydramnios. Its cause is unknown, but again, an association: such babies sometimes have no kidneys, or the tunnel from kidney out is closed. They can't pee, and their ocean dries up. They are wrinkled, leathery, dry; sometimes the amniotic drought melts the tissues, and the baby is born all clung to itself, webbed.

Too much, too little: both problems are found to cause immature lungs. Without the proper flow of fluid in and out, without the delicate interplay of urine and breath, the tiny sacs inside the lungs can't stretch, can't expand and strengthen for the coming air. The little lung alveoli are like imitation gills, in ready evolution.

All in the correct time the water splits the thin membrane apart and cascades out, from womb to earth like wave to shore—hot, sticky, humid, the amniotic fluid gushes down the moth-er's leg, soaks her clothes, and signals time. Deprived of liquor the fetus and its cozy home begin to move and labor starts, and hurries, till it ends in one great hungry rush of birth.

THE LIQUOR is always salty. Fetuses have salt-sensitive cells on their tongues, in their throats. What does the fetus taste? What colorful evocation appears to its memoryless mind? Salt as a nourishing bath in a dark, wet sky. A scientist tells that he offered samples of amniotic fluid to adults in a taste test. They called it salty. The scientist neglects to inform his readers if the participants knew what they were tasting. Did any of them feel an inexplicable tug of nostalgia, a remembered warmth? Would we have words with which to say?

3

METABOLISM

Manatees cry real tears, and seals sweat, and the urine of whales is thick with salt. Why should I be different from my fellows, mammals all, and all familiar with oceans of different kinds? It has been suggested that our kidneys are products of our marine origins, evolving the way fetal skin evolves, moving us from a free bath in the universal sea to separation—and motion. It is a lonesome sort of progress. Our blood and urine, the amniotic cradle like the bubble of Okeanos overhead, are cellular memories, a biological tradition.

The body knows its need for salt, and salt's power. The calm, unquestioned wisdoms are fertile and common: salt makes for a long life, for health, for fertility and potency. Salt makes you slim, repels gophers, bonds friendships. Salt water snorted out the nose is said to cure asthma. Salt is power, best harnessed with respect. Aristotle, man on the scene, suggested you could cure your elephant of insomnia by rubbing its shoulders with salt and olive oil. The Icelandic natives claim the first human was a block of salt, serendipitously licked into shape by an unwitting cow. Salt forms with humanity a conjunctive commensalism, a symbiosis in which we gain and it hardly notices our passage. Salt is necessary, compelling, an obligation: sewn secretly in the hem of a garment, salt cures homesickness in the one who wears it. Salt dances across cell walls, pads silently up the length of a muscle fiber, jumps a synapse, tickles the heart.

Salt softens the journey of sperm, buffering the harsh passage through vaginal acid. Three hundred million sperm in every ejaculate, swaddled in mucoid syrup rich with cholesterol and hormones, fats and fructose and salts—swimming due north. The smell of sperm is said to be one of the primary odors, unique in its damp sweet musk and the flavor of metallic salts it leaves on the tongue.

Tears are salt, drying on the cheeks in thin crystalline trails. In the tales there are tears of jewels, tiny rubies slipping down royal skin; tears of pearl and gold and blood; tears that break spells, make flowers grow; endless tears turning to rivers, springs, lakes, and the sea. A human cries about 4,000 times in its first two years of life, but it's bad to cry at the deathbed, say the tales—you "cry back the dying," make their leaving hard.

Leah cried for years, doomed to marry Esau against her desire, and the crying ruined her eyes: "And the eyes of Leah are tender, and Rachel was beautiful." Angels saw Isaac laid on an altar by Abraham, and cried angelic tears into his eyes, dimming his sight. In the old stories heroes search for tears lost in the ocean, sifting and sifting through pails and buckets and boats full of salty, warm sea.

And then there's blood.

THE HUMAN body can be viewed a lot of different ways. Chemists and biologists have their bias, and envision the body as a series of vessels laid end to end, one atop the other like bricks. The vessels are cells, and the cells are filled with salty water and the space outside the cells is filled with salty water and the space between the cells is filled with salty water. The cells—muscle, nerve, brain, blood, bone—are like gurgling sacs of protoplasm, breathing a salty fluid of changing composition as necessary as air to the lung. Sodium is the most common inorganic substance in the human body. Red blood cells require a dose of salt: too much and they shrink into mummy cells; too little, and the spongy walls burst with a skinny hiss, pop and pop.

A human adult is nearly two-thirds water (infants a boggy 80 percent) and all the water is murky, afloat with particles. A loss of less than a fifth of your body's water would likely kill you. A loss of a much smaller fraction of any of the various particles would as well.

Many of the flecks that thicken human blood are called electrolytes, substances that break apart in fluid and have an electrical charge. Vitamins, glucose, many nutrients don't dissociate this way. But minerals like salt slip apart into weakly charged ions of sodium and chloride, just as in seawater. They are measured not by weight, but by chemical power—the power to combine with another substance.

Electrolytes bathe and nourish cells in a continual dance of neutrality, each negative leaping to balance a positive, the differential between sparking tiny slices of power here, then there. Electrolytes need to scoot about from space to space, blood moving into kidneys, plasma moving into cells then down a vein. With every motion of an ion water follows, keeping cells padded equally inside and out.

The world likes to keep things even—if not equal—and has laid out a set of rules toward that end. One such basic principle is osmosis, the passage of solutions across a membrane until their concentrations balance. Osmosis is one of the great motive forces of the body, ever at work—the flow of water from space to space judged by the changing congress of sodium and calcium, potassium and chloride, their size and power and push. Sodium is the dominant cation, or positive particle, in all these transactions, leading a dip and swing of synchronicity from cell to cell.

Every ion of sodium is pregnant with an unbalanced charge, a shivering incalculation of spinning electrons tumbling from shell to shell all at once in dreamy quanta. Positively charged ions like sodium mill against the outside wall of a cell like a hungry crowd, drawn by the negative anions like chloride

pushing against the wall from inside. And sooner or later in this millisecond universe something happens—a change in temperature, pressure, a shock or flood of nutrients or toxins, damage or the sudden hormonal demand of fright or lust or anger—and the wall crumbles like Jericho's. Doors open, sodium rushes in, hangs around for half a trice and runs out again. One big exciting mob of electricity racing through, and the message is flashed along a nerve body as though on a telegraph line. Muscles contract, the heart pumps, a hand cups a breast, mouths meet and salt passes through, tongue to tongue. Salt burns wounds with an excruciating neural excitement, just the same—and in all this sodium isn't changed. It simply moves, flit flit, slippery thing: now you see it, now it's gone.

Chloride, on the other hand, is quiet, like the straight man to the comic. Chloride plays its role in osmotic pressure, as an anion, or negatively charged ion. It donates power to muscles and nerves, helps make enzymes, argues with bicarbonate over sodium in little spats to keep alkali from overwhelming the cell. Chloride is best known in the digestive juices as a part of hydrochloric acid, the ulcer's bane and antacid's target—but without the chloric in our stomachs, food would decay like a heap of peelings put out with the garbage.

As a person grows the amounts of sodium and chloride in his body change. The gross amounts increase, as body weight increases, but the amount per pound diminishes. A pound of infant flesh has nearly two-thirds more sodium and twice as much chloride as a pound of adult flesh. Altogether a 70-kilogram (155-pound) man has about 130 grams of sodium in his body, and needs less than 3 grams a day to maintain it. (Don't let it stop you; the average American adult eats between 8 and 10 grams; a Japanese as much as 25 grams of sodium a day. Those most remarkable filters, the kidneys, can retain and release widely varying amounts of sodium to keep the books balanced.)

In fact, the fabled "relatively healthy adult male" does all right for himself with sodium. He generally eats what he pleases, and is luckily pleased by a fairly nutritious diet, rounded off by chips and pretzels, pizza and beer. His sodium load goes up and down, hour by hour and day by day, as does mine and yours. The busy kidneys compensate, typically filtering through nearly 500 grams of sodium a day (the same ions, the same grams, round and round), but excreting fewer than 5. The healthy adult female, on the other hand, retains extra sodium before each period of menstruation, storing up for a potential pregnancy. If she gets pregnant, she will eventually accumulate an extra 10 to 20 grams of sodium in her body, stored in the amniotic fluid and the fetal plasma. Both men and women lean on their kidneys, basking unconsciously in the renal accounting of ions.

Salt burns wounds, and revives burns. It is always on a budget, the body's salt, and in times of crisis salt can kill. A bad burn bleeds the precious fluids away like a faucet, pouring out proteins and nutrients and salt in a flood; pour the salt back in proper measure and the leak slows, checked by the salt's osmotic virtue. But who pays attention when the figures match, accounts balance? Have another pretzel, my friend, and listen to the tales of disparity, when the scales tilt.

The most common cause of hypernatremia, or an excess of sodium in the blood serum, is dehydration. Other causes include kidney failure, diabetes, cancer, the desperate drinking of seawater, the accidental mix of salt in baby formula. With too much sodium the fluid outside the cells expands and the fluid inside contracts, shrinking the body cells, throwing electrical charges into chaos. The pressure in the brain first shoots up, then drops, brain cells contract with a tight sibilation like a bicycle tire losing air, blood cells burst. Hypernatremia is hard and slow to treat, requiring delicate juggling of water and sugar and diuretics, chemicals that flush water and minerals from the body.

Hypernatremia is studied, primarily with animals, for clues about kidney function, high blood pressure, human endurance. Too much salt isn't always a bad thing. One of the more intriguingly brutal experiments of recent years involved dogs. The experimental group were fed high-salt diets for a spell, while the controls ate a normal diet. All were then bled nearly dry, their femoral arteries sliced open, bled until the dogs fell into shock and their blood pressures plummeted. Then each dog was transfused with its own blood again (the hole since stopped up), and when the study was over, the dogs on the high-salt diet had a better chance at survival. Not, perhaps, a circumstance we are likely to find ourselves in very often, but one to keep in mind.

It's not all dogs, either. People submit, and the descriptions in the literature are as dry as the mouths of the human guinea pigs involved. A radiology journal recently published a terse, detailed study of the radiographic—or X-ray—changes seen in the chests of men after they were salt-loaded. Radiologists often search chest X rays for evidence of fluid in the lungs, or an increase in the size of the heart or the heart lining. Such findings bode ill, and dictate serious treatment. The designers of the study wondered if a person filled to the gills with salt could mimic, on film, such chronic heart diseases. This is how they found out.

Eight "healthy male volunteers"—we aren't given a clue as to why a person would volunteer for such an ordeal, or how much each was paid for his trouble—entered the research unit and were thoroughly examined, blood and urine and bone. Then they were fed, for a few days, on a diet containing a mere 10 milliequivalents of sodium: scarcely a crystal. Over the next several days, the salt in their diet was gradually increased, the food growing heavier and heavier with brine, their bodies accumulating more and more sodium. By the last day, the men were receiving 1,500 milliequivalents, equal to thirty-four grams of sodium; they slept that night with an intravenous line pumping salt solution into their parched veins.

"No ill effects were observed," write the researchers, and then mention, in passing, the swollen feet, and that each man gained an average of eleven pounds. The X rays showed hearts remarkably swollen, surrounded by swollen, stretched blood vessels. Blood pressures rose, the strength of the heartbeat and the amount of blood it pumped rose, too. The hearts were working hard; the organs looked like the organs of old, long-sick men. How did they feel?—I can't help but wonder. Thirsty and dry, hearts pounding, skipping a beat, skin and lips and tongue dry and cracked? Was it like lying too long in the sun, like a trip across a desert, like the rush of dry heat from a stoked furnace? Do we suppose each man gradually deflated, collected a check, and drove home, dreaming of a long, cold beer? It's like pickling in the Dead Sea, dreaming of a spring shower.

There is one bizarre cause of excess sodium in the body, uncommon but remarkable. It is iatrogenic, or physician-caused, and combines the mundane with the strange, leaving sodium as a footnote.

The occasional unfortunate grows a bezoar, an entangled mass in the stomach or intestine, composed mainly of decayed food and hair. Bezoars are sometimes found in the stomachs of animals, and were for a long time believed to have curative properties. In the Far East bezoars were thought to be magic stones good for healing; in old Europe royals would keep bezoars in jeweled boxes by the bedside. (The peasants could rent one, when necessary, from apothecaries.) Bezoars can grow to be several pounds in weight and cause difficulties for the bearer. In times past they had to be cut out. Then Adolph's meat tenderizer came along.

Meat tenderizer contains papain, a protein-digesting enzyme, effective both in breaking down the muscle mass in a steak and dissolving the proteinaceous bezoar. Typically the powdered tenderizer is made into solution and then washed repeatedly through the digestive system until the bezoar lifts. This seemingly perfect solution, though, is being abandoned in favor of the knife again. It has two side effects: sometimes the papain dissolves the muscles of the throat. And because it is so high in sodium, the occasional patient has suffered from a sudden and frightful overdose of salt.

A LACK OF SODIUM, or hyponatremia, can be no less dramatic. Too much water, too little salt in the diet, can cause it. Or a different kind of kidney failure. Usually when the kidneys give out they hang onto sodium, thickening the blood. (A simple defect in excretion that allows only one milliequivalent a day to be reabsorbed—23 milligrams—would create a concomitant flood of two extra liters of fluid in the body in just a year.) But now and then the kidneys flush too fast, pouring the sodium (and more, protein and sugar and nourishment) out in the urine, a wash of straw-yellow liquid full of vitality, down the drain. Hyponatremia can also be caused by diarrhea and vomiting, cardiac failure, and more.

Either way, the body's electrical potential fades, like old ink. The cells are thirsty and malnourished, the flow of urine slows to a dribble. The victim feels apathetic, lethargic, confused; he or she acts strangely, overestimates the passage of time, complains of cramps and shows poor judgment. One anthropologist, noting that certain American Indian tribes limited salt intake for warriors before a battle, suggested it was the symptoms of hyponatremia that were sought. Such a person, he muses, "would make a formidable warrior"—rash, wild, incautious.

A DDISON's disease is a defect of the adrenal glands which, among other things, causes a lack of aldosterone. Aldosterone is a hormone that causes increased salt retention; without it, the body excretes far too much sodium, holding tight to its potassium like a child to an all-day sucker, until enormous and bizarre imbalances result. Addison's victims are bronzed with an eternal tan and splotched with bluish stains and black freckles. They are thin, lethargic, anorectic, and cold. Their hearts weaken and eventually shrink. Addison's patients develop intense and uncontrollable hunger for salt, searching for it in nooks and crannies, sprinkling it on every bite, mixing salt in water and drinking it like soda.

In 1940 a 3½-year-old boy was admitted to a children's hospital. He had almost fully developed sexual organs, pubic hair, a deep voice, pigmented nipples and a brownish tinge to his skin. "In the hospital the boy did not seem to be especially ill," writes an attending doctor. "He behaved like a very defective child, growling and snarling in an incoherent manner when attempts were made to examine him. He was offered the regular ward diet. His appetite was poor and he ate but little of the food. When feedings were forced, he vomited on several occasions." The physicians were puzzled by the case. Seven days later the boy suddenly died.

Only after an autopsy revealed greatly enlarged adrenal glands did the truth of the matter surface. The parents stated that since the age of 12 months the boy had eaten salt with a voracious longing, often unable to hold down any solid food.

"When he was about a year old he started licking all the salt off the crackers and always asked for more," stated the parents in a letter to the physicians after the boy's death. Eventually the boy began chewing the crackers for the salt, spitting out the mush that resulted, and did the same with bacon, eating, wrote his parents, "a terrible lot" of salt.

Soon the boy learned the secret of the salt shaker, and refused to eat unsalted food. He would cry and point and pout until the distraught parents (who had been assured by the boy's doctor that it was fine to indulge his craving) would cave in. Salt was one of his first words; he discovered olives and pickles and pretzels and potato chips. By the age of three the boy was able to eat solid foods without vomiting, provided the salt content was high, and had begun his precocious and disturbing development. He disliked sweets, couldn't eat cereal if it had sugar and milk on it; he ate Wheaties dry with salt sprinkled on. The boy's life focused on his condiment.

"When he was two and a half he loved to hunt for the attractive pictures of all the good foods in the magazines. He learned to know what all the foods were and made believe he ate them," wrote his mother in the letter. "There was no other one food he seemed to crave like salt, except water," which he preferred to milk. "And when he saw the river or the ocean, he always thought he had to have some to drink . . ."

The sad conclusion to the life of this ocean-borne baby was a sudden death, separated from the parents who had tried so hard to please him, fed the nauseatingly saltless diet of a hospital ward, battling the strange hands with incoherent growls, begging salt. He died of hyponatremia, salt loss, his adrenal glands wildly awry and pouring out the hormones that hold in the salt.

THEN THERE is water intoxication, internal storm, the body drunk with fluid, inebriate. The volume of water both inside and outside the cells increases, but the salt does not, and brain cells swell, then shrink. (All those precious memories up there, elastic in the wind of shooting electrons.) Water intoxication can occur accidentally, especially in the medical treatment of a dehydrated person. But it happens most frequently among schizophrenics.

For reasons as mysterious as any other in the realm of psychosis, schizophrenics sometimes have a compulsive need to drink water. Compulsions are exactly that, impossible to overcome, demanding in the most lordly and imperative of voices. Compulsive water drinkers, if denied a cup and tap, will drink from pans, aquariums, toilets, flower vases, radiators. Patients have been known to drink 35 liters of water a day, and their bodies swell like sponges. They are plunged into severe hyponatremia, the sodium ions scant and embarrassed in all that ocean. Water intoxication is deadly, and the usual treatment pitiful and cruel, exquisitely simple: lock the driven person in a dry room with his demon, and let him cry his possession out. Or hook up a vein with a flow of salty, salty fiuid, and try to keep up. One psychiatrist describes treating four schizophrenic water drinkers with oral salt tablets, hoping to match them drop by drop and keep the body balanced in some small way. He fed his patients gram after gram; one ate 62 grams of sodium every day. The treatment seems to work; the patients live, their compulsion is satisfied, if not cured. And only two of these pickled people had a rise in their blood pressure.

(A psychiatric aside: Manic-depressive patients are dependent on lithium to control their symptoms, but lithium is a tricky drug. Patients frequently get too much in their bloodstream and it starts to poison them. Lithium toxicity can be treated with a flood of salt water in the veins, which speeds up the body's elimination of the lithium. Then, of course, the person's overdose of salt must be treated.)

Sometimes, when fluid is lost—especially in vomiting, when large amounts of hydrochloric acid are expelled in that sharp unpleasant tang of puke—too much chloride is lost as well. Then the body's sodium combines with a base to make sodium bicarbonate, alkaline, upsetting the delicate pH of the fluids. Too much and the victim enters a state of alkalosis, straining against tetanic muscles and a delirious mind, nerves in twitch and heart unnaturally alert. Too much chloride, though, sets up acidosis, and the person suffers a vague lassitude, nausea, breathes rapidly and deep. Treat the first with a solution of salty water; treat the second with the antidote—sodium bicarbonate.

A rare disease of children is called Bartter's syndrome; the defect appears to be an inability of the kidney to hold on to the body's salt. Sodium and chloride slip away, potassium is lost in compensation, the patient suffers multiple and bizarre confusions of metabolism. Children fail to grow, their muscles dance and cramp and refuse to obey. Enough salt to pay the debt, and the symptoms ease—a temporary cessation. It's no cure—who knows the syndrome's origins?—and damage is done, to the kidneys and the adrenal glands. But extra salt makes a difference.

Cystic fibrosis, congenital and incurable and always fatal, was long thought to be a defect in sodium use. But new research suggests it is a defect in chloride, and the strange tricks of sodium—the heavy loss of sodium in sweat, so that a loving parent kisses her baby and tastes salt—follow. It seems that the chloride ions get stuck, can't cross the cell walls to help balance water, and somehow cystic fibrosis in all its mucky, mucoid congestion is born.

One more word on chloride. In 1980 Neo-Mull-Soy, a brand of soy-based baby formula, gained 20 percent of the market. Within a few months 34 cases of chloride deficiency in infants were reported, appearing as metabolic alkalosis, usually extremely rare. The babies were losing weight, lethargic, stunted. All but five of them were fed Neo-Mull-Soy. Neo-Mull-Soy is very low in chloride, a mineral long considered of minimal importance in babies because of its ubiquity. But the trend to low-salt diets and the public-relations ploy of selling "no-salt-added" baby food to interest the nutrition-conscious had created a new demand—a metabolic demand, a chloride hunger. Switched to other formulas, the babies seemed to recover. Neo-Mull-Soy changed its ionic equation, and chloride slipped back into a kind of dietary transparency: there, necessary, ignored.

I MET A woman, an old Jewish woman living in a nursing home, who was terrified of salt. She thought it would sprinkle from the ceiling and pickle her skin. She was afraid, when she left her room for a meal, that someone would come and spray it with seawater and brine. She could smell it everywhere, feel its grit on her arms, its dry dust in her nose and mouth. She studied the Old Testament, her feet growing cold, feeling more rooted and unable to run with every passing day.

Much as I understand such a burden of regret, such iconic concern, somewhere in the midst of these last few years of study, salt and I became friends. (Not that we weren't already acquainted; we'd spent a lot of time together.) But I worried about salt, especially sodium, the noisiest of the pair. I read these strange stories of dogs bled like medieval witches and acquiescent men puffed up like adders with salt till their blood runs like dirt, like a salty mudflow—and I worry. I learned strange things: Alcohol increases the hunger for salt. A surprising number of people who have cancers of the digestive system—pancreas, stomach, colon—love salty food, excrete more sodium than normal. In rats salt makes tumors grow. But which is cause and effect? Does alcohol make you crave salt, or does an abnormal craving for salt make you thirsty for the drink? Does salt cause cancer—or does cancer have an appetite for salt?

Well, I don't fret much. I've got room for salt's sharp sweetness, the fine and delicate coat disappearing with each hot bite. My blood moves over, my arteries sigh and open with satisfaction. Hurt me? Not salt—not me. I'm a fan. I've got salt all over me inside and out; salt is spread out across the world, and little parts of it take turns residing in me. I can't lose salt even if I try; it hitchhikes along, a barnacle in my blood.

4

URINE

GOD MADE the pig from urine, the tales say. Urine can melt
rocks, exorcise witches, create floods, and attract butterflies.
A child's urine is said to turn gold to ashes, and the peculiar
smell of it reveal a man's true origins.

Every human lifespan has its moments of unadulterated
pleasure, and the sudden gushing release of a long-held flood
of urine is one we all share. Babies revel in their high-flying
streams, golden fountains squirting to heaven and bursting in
a cascade warm as the body's middle. Every mammal pees—
whales pee super-salty thick flows to trail behind them like
thin salt currents in their wake. My cat marks her annoyance
at my slight transgressions on a paper I left carelessly about;
when I find it the next day the dark ring is painted in glitter,
tiny crystals of dry urine like something I might find on a
sorcerer's spice rack, pheremonal and strong.

The premammalian creature, our ocean ancestor, developed a simple kind of kidney, not only to eliminate wastes but
to maintain the creature's osmotic balance in the saline waters.
When vertebrates began to crawl up streams into fresh water,
their needs changed. Salt preservation and elimination became
crucial, and the conservation of water essential. Parts of the
mammal kidney—remarkably similar in structure between
species—are found in fresh- and saltwater fish today. With
time (small eons, atomic time) the kidney of the modern mammal asserted itself, through the trial and error of evolution, to

a premier piece of work. The modern kidney may be only a precursor of something even better, but it is already a filter of exquisite detail and efficiency, essential, the perfect solution to a strange and incomparable problem.

The little kidney, hanging with a certain itchy vulnerability beside the spine, is a heap of entanglement, mazes within mazes, zigs within zags. Its working unit is called the nephron, microscopic in size and twisted upon itself like a miniature plate of spaghetti. Every kidney has about one million nephrons, each fed and drained by tiny blood vessels called arterioles. Blood plasma—blood emptied of its red blood cells—knocks against the nephrons endlessly, in such quantity that two normal adult kidneys filter about 180 liters of plasma every day, an amount equal to the entire extracellular fluid volume every two hours.

Plasma enters the nephron through a port called the glomerulus, a convoluted membrane that (in health) passes water and electrolytes freely and blocks the passage of larger molecules, like protein. Plasma turns along this road into glomerular filtrate, 94 percent water and 6 percent solutes, including sodium, chloride, calcium, magnesium, potassium, organic acids, and more. Out the glomerulus pours a steady stream of filtrate into the proximal tubule, into one million proximal tubules, each so labyrinthine that, stretched straight, one would be half the width of the kidney itself. The proximal tubule is lined with uncountable numbers of villi, projections waving in the stream like stalks of wheat in the wind, each a part of the surface area of the tubule itself. More than half of the water and sodium that will be reabsorbed from the filtrate back into the body's circulation will be removed here, along with amino acids, glucose, potassium, and other substances, at rates determined by the amounts of each in the blood and body tissues. By the time the liquid starts to cascade into the narrow hairpin of the loop of Henle its character has changed, becoming—but not yet—urine, unique, sterile but full of waste.

The filtrate slides through the loop and crawls out the distal tubule, a cousin to the proximal, and by then 99 percent of the original glomerular filtrate has been reabsorbed. The wiggling villi hairs have sucked it of sugars and salts, hormones and amino acids, and water. What is left after these few dozen torturous centimeters is urine: transparent, slightly acidic, with a bitter saline taste, 95 percent water and the balance composed of urea, ammonia, acids, a blood constituent called creatinine and—in health—very little else.

The kidney and salt are mutually dependent in multilayered ways, but the mechanisms by which the two connect are still poorly understood. It is one thing to draw a pressure gradient from charts of oncotic and osmotic pressures and filtration rates and predict the amount of sodium held here, released there, and quite another to know how such things occur—to know *why*. Each new clue seems to contradict, rather than illuminate, previous clues. Every conclusion seems an untoward assumption.

The kidney has a most intelligent set of answers to environmental conundrums. If I drink a large amount of fluid, my bladder will fill with the amount of urine necessary for balance in about 45 minutes. If I stop drinking for a spell—as when I sleep—my urine grows more concentrated, darker, stronger-smelling. (The actual production of urine itself slows during sleep, but it never stops.) If the blood presented to the glomerulus is lacking in something particular, like potassium, the amount of potassium excreted simply decreases through a complex but predictable series of changes in pressure patterns and membrane permeability.

The normal kidney handles about 25,000 milliequivalents of sodium every 24 hours, excreting less than one percent. But the amount excreted depends upon the amount taken in, and a healthy kidney can handle a sodium intake between 1 and 20 grams a day. Less in, less out. Too much sodium, then more sodium passes into urine. Simple.

There is some correlation between high blood pressure and the amount of sodium lost in the urine, but conclusions are tricky. This is no different from the alcoholic salt-eater: chicken or egg? It is too easy to claim the kidneys compensate for an overabundance, puffing in the effort, because the amount of liquid pushed into the glomerulus is largely a result of the pressure exerted by the blood in the first place. Perhaps a person with high blood pressure has a sour glomerulus, and the increase in pressure is an effort to compensate for a stubborn membrane.

The end result of these incalculable miles of tissue (such lovely names: juxtoglomerular apparatus, minor calyx, urethral meatus, the bulbourethral gland, the ever-present macula densa) is urine. A singing poem of excess ocean, clear and bright in the noonday sun, as sterile in its perfection as a surgeon's forceps.

It's good for things besides the demarcation of territory and the glorious release of a filled bladder. An egg boiled in the first morning urine of a person ill with fever, before the urine grows cold, is said to break the fever. The act of making water, done in certain places and circumstances, can work magic. Women who feared they were pregnant urinated on a freshly thrown molehill. (Even now a hormone found in the urine is the most common test for pregnancy. A few drops of pungent fluid stirred up with the right chemicals for a few moments tells the story. If it grows grainy, no baby. Smooth as silk, less than nine months to go.) Children who suffered bedwetting used to be taken to a graveyard and told to urinate on the grave of a child of the opposite sex—an experience likely to tighten the most stubborn bladder.

Sooner or later that sphincter has to tighten. One can't go around happily percolating urine for all the world to see. So we train our children to suffer the small pangs of a distended bladder, to "hold it," take control, delay that gratification for the proper moment of privacy. Every language has its formalities, the little polite euphemisms for toileting—and every language has its slang, too, the gleeful words of a slightly prurient pleasure in pissing. It is a measure of intimacy, how easily we relieve ourselves in front of another, how much we can relax that tensed muscle.

Sad, then, to lose the right, to have one's bladder fail to respond. All manner of inelegant problems can arise: spasms and tumors and swellings of various kinds, unnatural twists and constrictions, pockets leading nowhere, deadened nerves, pathological anxiety. Then the bladder fills but doesn't empty— fills and fills till the lower belly grows round and drum-hard. It can hold more than a quart if necessary, but few of us could happily tolerate such a load. Such a situation can't be ignored. If you can't pee, then someone has to do the job for you with a catheter tube.

Kidneys fail, too, for a hundred reasons and more: drug allergies, cancer, diabetes, high blood pressure (which can be both cause and effect), heart disease, massive infection, poisoning, congenital defects. The kidney forms stones, beautiful crystalline seeds growing in saturated urine, made of calcium and oxalate and phosphate, with pretty names like staghorn and sand calculi.

When the kidneys fail, all that magnificent, unsung work of filtering must be done by something else—a dialysis machine. Dialysis is the process of pushing a solution through a membrane. People on dialysis are hooked, artery and vein, to an electric membrane several times a week for several hours at a time. If they don't receive this treatment they die, rapidly, from the accumulation of wastes in the blood. The dialysis patient almost always has to have a carefully controlled diet, counting each spoonful and sip, a diet often mundane in the extreme. Typically such a menu is high in calories, often low in protein, with every drop of fluid measured. In severe kidney failure, if the kidneys are hanging on to the precious sodium and potassium without reason—confused, unable to read the signals—the diet will be very low in sodium and potassium as well.

Kidney failure causes chronic nausea and a loss of appetite as well, so getting the calories in is a trick. Some patients make little balls of butter rolled in sugar for snacks. And no *salt*, heaven help us, no salt: no chips, pickles, pretzels, jerky, sauerkraut, buttermilk, popcorn, cottage cheese, bacon, olives, canned soup, peanuts, ham; no pizza, potato salad, TV dinners, breakfast cereal, salad dressings, sausage, corned beef, bouillon, eggs, and more. (Potassium is still another story.)

Some patients are salt wasters, flushing the minerals and electrolytes out of their bodies as fast as they can shovel them in. Such patients may need as much as 20 grams of sodium every day, twice what we typically eat. I've seen patients on dialysis eating plates full of french fries while every other person in the room stared with a kind of blank and hungry envy: at the salt-spillers, the luck-wasters, squirting out their streams of billowing, singular, pelagic waters—becoming, in their extravagance, one more small current in the waves.

5

BLOOD PRESSURE

HIGH BLOOD pressure, or hypertension, is a condition granted
considerable metaphorical power: we surround the word with
a host of contextual meaning. Hypertension is tight, tough,
rigid, and strict; it is fast-paced, tired, hungry, a bit unhappy
and worn. It inhabits middle age, three-piece suits, and soft,
office-broken bodies, mostly male, down in the mouth, quick
to temper. High blood pressure is a jam-packed cultural fault,
a slang of life-style and desire, a debt being paid. It is a little
bit moral—you asked for it, buddy.

My medical dictionary calls it "tension that is greater than
normal" before defining its myriad forms, and already we're
running in the dark. I have been asked many times what a
"normal" blood pressure would be; the question usually comes
sideways, casual, as I am puffing up the cuff on a worried
man's arm. I batten down the balloon over the pulsing veins,
and the whole bleak future battens down, too.

First, a definition. Blood pressure doesn't measure the
pressure of blood. It is pure physics: it measures the pressure
blood exerts in its passage, and from there is a clue to the
elasticity of the blood vessels themselves. A healthy artery is
a fine compromise between strength and stretch; it must carry
a constant pounding flood of blood for—one hopes—a long
life, but it must bend in the changing waves, pliant and willing
to roll. The pressure is measured indirectly most of the time.
(The first proof that blood circulated under pressure came in

| 37 |

1733, when Stephen Hales stuck a nine-foot tube in the carotid artery of a living horse and saw the blood rush up and start to beat, up and down, inside it. This was the first *direct* measurement—crude but effective.) To measure blood pressure indirectly one tightens a cuff around the upper arm (or thigh, if the arm is unavailable for some reason) until the pressure in the cuff exceeds the pressure of the heart's pumping in the brachial artery, and the artery collapses. Then the pressure in the cuff is slowly released while the nurse or technician watches the gauge or dial and listens through a stethoscope for the arterial pulse to return.

The first sound heard is called the systolic pressure, and measures how much resistance the large blood vessels near the heart have to the pumping of blood—in other words, how elastic they are, or how rigid. The second number is the diastolic pressure, counted when the pulsing sound in the stethoscope disappears. (Trained ears can hear three, four, and even five sounds sometimes, each with a slightly more refined meaning.) The diastolic pressure is a measure of how tense the blood vessels are in the moments between heartbeats, when the heart relaxes—that is, their softness and pliability. Because the pressure gauges are calibrated against a standard based on the pressure required to raise a level of mercury against gravity, blood pressure is measured in millimeters of mercury, or mm Hg. Pure physics, really: if the volume moving through a pipe increases, the pressure in the pipe increases. And if the width of the pipe decreases but the volume moving through it doesn't, again the pressure increases. So extra blood or fluid can cause high blood pressure, and anything narrowing the space or reducing the stretch of the blood vessels also raises blood pressure. The force that must be exerted to circulate the blood increases, and little by little by little, every cell in the body pays the price.

There is no such thing as a normal blood pressure. Each of us is a body in motion, a set of precise and mysterious patterns weaving and unraveling continuously. Blood pressure is a simple marker for a series of balancing acts being performed by acrobatic systems of hormone and enzyme and electrolyte; it drops when we sleep, rises when we run, eases when we are relaxed and shoots heavenward when we panic. High blood pressure is notoriously silent: except for the simple measurement, it may have no symptoms at all until damage is done. Of the 30 million or so Americans believed to have high blood pressure, about 5 million are ignorant of their condition.

The medical establishment generally holds that hypertension is systolic blood pressure persistently greater than 140, or (and sometimes and) diastolic pressure greater than 90. It is best measured several times in a row—because, oh patient, there is nothing like sitting starkers on a table watching that needle bounce upward to excite the arteries.

It is worth worrying over. A hypertensive person is at considerable risk for heart failure and kidney failure and that great eraser of the brain and its pleasures, stroke. It kills, leaving great swaths of damage on its way. And of all the cases of hypertension in the United States, about 90 percent are of unknown cause.

I have long seen the work of high blood pressure in my father's face. It is partly the force of blood cells squeezing into tiny tightened capillaries, and partly the force of personality. His is a face of flaccid tension, red and hot and winded but soft all at once—a sprung strength. He is beaten down and full of vitality too, wound up a notch, a 45-rpm man. His blood pressure is sped up like his rage and the curlicues of bright ideas he unravels, bursting explosively, suddenly, insidiously upward till his eyes glisten with the pounding of the heart behind them.

So, will it be you or me, brother? Perhaps both. Hypertension plays favorites. Blacks of both sexes are the most susceptible, and among whites, women are more likely to have it, regardless of weight. A diet high in meats and cholesterols seems to increase the chance of developing high blood pressure. Defects in circulation, hormones, and kidney function are all associated with high blood pressure, though not clearly as cause or effect. Inexplicably, blood pressure in infancy seems to be a fairly good predictor of blood pressure in later years. Pregnancy and oral contraceptives increase blood pressure. Once the blood pressure begins to rise, at whatever age, it has a powerful tendency to continue climbing of its own accord, in spite of drugs, diets, and exercise. Prevention seems best— but no one agrees on what works. Genetics may play the most important part, the bugaboo heredity. But always the conversation returns to salt, a simple answer, a thing to pinpoint, mild and common and desired.

A potent determinant of sodium transport in the body is a hormone called aldosterone. Aldosterone is secreted by the adrenal glands and causes the kidneys to reabsorb more sodium before it passes out of the body. To balance the electric potential in the fluid aldosterone causes a concomitant loss of potassium. (Neat trick, that.) Aldosterone may affect the amount of sodium in the saliva, and somehow or other it certainly affects our craving.

Aldosterone is only the end product of a series of reactions, thought to begin—as such things can begin, only by entering a circle at some point on the line—with a substance called renin. Renin acts as an enzyme on angiotensinogen, to create angiotensin I, a compound of ten amino acids. Then a converting enzyme takes over, slicing off two acids to form angiotensin II. Angiotensin II itself is a powerful pressor hormone— a substance that directly raises blood pressure. It is also the major stimulant for secretion of aldosterone. Retention of sodium *may* cause high blood pressure, so, walking the circle of cause and effect backward, increasing levels of aldosterone are important in that consideration. Since anything that stimulates renin, like changes in artery pressure—or emotional stress—eventually stimulates aldosterone, the entire renin system becomes important. But this is just one of many cycles at work, patterns that roll, swell up, slide back like tides. To say

salt causes high blood pressure is so simple as to become
fallacious.

This implication of salt as a factor in blood pressure began
a long time ago; by 1925 a physician was *refuting* the theory
based on the most recent evidence. The research since has
gone hither and yon, into smaller and smaller pieces, details
of detail, subthemes in plot devices. It is almost a self-repli-
cating machine, this hypertension establishment, limited in
scope and repetitive, most studies centered on a few subjects
and barely stepping ahead. Almost all the research (and this
is built into the nature of allopathic research) moves a bare
step ahead with each result. Few of the published studies
attempt to place results in a large context. As they rely on rapid
and erratic biochemical changes in genetically programmed
animals in laboratories, that isn't so surprising.

"Altered Arterial Muscle Ion Transport Mechanism in the
Spontaneously Hypertensive Rat." "High Salt Intake Blunts
Plasma Catecholamine and Renin Responses to Exercise: Less
Suppressive Epinephrine in Borderline Essential Hyperten-
sion." My favorite for the week: "Blood Pressure, Intraeryth-
rocyte Content, and Transmembrane Fluxes of Sodium During
Normal and High Salt Intake in Subjects With and Without
a Family History of Hypertension: Evidence Against a Sodium
Transport Inhibitor." All those capitals! All those molecular
clues, those electric bursts of inspiration and shifts in focus,
all that data! And still, does salt cause high blood pressure?
Well, yes and no. Or rather, maybe. Or, we're not sure yet.
"We are confronted with a bewildering (and perhaps exces-
sive) array of observations," writes one of the leading researchers in the field, "including many apparently contradictory ones."
Yes, it appears that salt—more specifically, though not surely,
sodium—does in some way elicit, or initiate, or permit the
development of hypertension. But what of chloride, potassium,
and calcium? What of congenital defects in electrolyte metabolism? What of the interaction of the many substances in-

volved? And what of the possibility that the noted differences between hypertensives (you, perhaps?) and normotensives (like me, lucky me) are simply artifacts, red herrings, shadows thrown by the real cause? What of that?

Anthropologists have described more than 20 distinct cultures that, among other traits that place them far from American society, have normal blood pressure throughout the life span. We tend to assume our blood pressure will rise with age, but in these cultures it does not; the blood pressure of a young man will change little until he dies of old age.

The Yanomamo Indians live in the rain forests of northern Brazil and southern Venezuela. There are about 12,000 to 15,000 Yanomamos, and they dwell in small, isolated villages having little contact with each other. Until the 1950s, when missionaries began to establish small contact stations, the Yanomamos had no continuous contact with Caucasians or Westerners. Their primary source of food is a large banana used in cooking—supplemented by game, fish, vegetables on an irregular basis. The Yanomamos have never eaten sodium chloride; in fact, they had never seen it until it was introduced a few decades ago by outsiders.

The Yanomamos were studied at length in the sixties, when they were first in the hands of Western medical scientists. Surprises, many surprises.

The Yanomamos show an increase in blood pressure from childhood to adulthood. But after that their blood pressure stays the same, and in some cases declines. The mean blood pressure for men over fifty years of age was 100/60, lower than mine has been for years. But when the researchers studied the amount of sodium excreted in the urine, they found that the Yanomamos lost less sodium than ever before recorded. And when their diet was examined, they were found to have the lowest intake of sodium ever recorded, a level that had been thought to be incompatible with life. They are seldom overweight, and don't gain weight with age. They are physically active throughout the life span, and have few health complaints.

The Western world encroaches, the rain forests fall, and salt is a commodity. The banana will come to seem mundane. The children will complain, the grandparents shake their heads. Those who go to the town to discover its riches will return tenser, fatter, out of shape, with tales of an easier life. I am only guessing, of course. But it has happened a number of times already, to others.

The Yanomamo are unique only in the extremes of their metabolic range. The same lifelong health—low blood pressure, virtually no heart problems, little obesity—has been found many times in rural cultures. The Papuan people of New Guinea, who eat a low-protein, low-salt diet with more than 90 percent of their calories coming from carbohydrates, also show no signs of diabetes or gout.

It is easy, considering all the supporting evidence, to dismiss such biological wonders as the product of genes—the gift of a capricious God to an otherwise unprivileged people. But of the twenty-odd cultures that have presented this picture, most have yielded to the pressure of westernization— and given up their health in exchange. The Chimbu Melanesians of New Guinea and another Melanese tribe on Cook Island, the Nagovisi of the Solomon Islands, and the Samoans—tribe after tribe has assimilated into Western culture and grown ill. The straightforward change of environment, country to city, results in many more subtle changes. The growth rate of children increases, and the size of newborns increases, too—possibly due to more protein—and therefore the number of birth complications rises. The average percent of body fat increases, and the ability to conserve oxygen drops. Longevity actually increases but is accompanied, for the first time, by an undeniable rate of sudden death in middle age. And even as people live more years, their biological fitness diminishes, and chronic, debilitating diseases appear.

A number of things are at work here. The diet changes, suddenly, irrevocably. The environment offers new and unexpected stresses, as obvious as the psychological stress of assimilation and as disguised as the lack of physical exertion. The shock of assimilation alone has been proposed as the cause of the gradual but dramatic climb in blood pressure shown by such groups. But the diet changes, too: fat, sugar, meat consumption goes up. Salt appears, everywhere. Obesity is suddenly a problem.

Most of the undeveloped cultures depend on fresh foods, and eat a diet high in potassium and naturally low in sodium. (Two exceptions. The Qash'gai nomads of Iran eat a lot of salt and get high blood pressure; a group of Nepalese villagers that import salt illegally from Tibet in Himalayan mule trains eat large quantities of salt and *don't* develop high blood pressure.) The inescapable fact remains that cultural groups who traditionally eat little salt and maintain normal blood pressures develop hypertension for the first time when they add salt to their diet.

The salt in drinking water has been correlated with lifelong increases in blood pressure, no matter that no one understands the mechanism. A number of studies of school children have shown that people living in areas with a high-sodium content in their drinking water have a significantly greater risk of high blood pressure, beginning in early childhood. The American Heart Association recommends that drinking water have less than 20 milligrams in a liter for those on low-salt diets; the state of Massachusetts has identified 106 towns with water supplies exceeding the AHA limits. Water drawn from the Ogallala aquifer, the underground water feeding much of the Great Plains and Southwest, has almost 1,600 milligrams per liter in places.

Unfortunately—for us simpletons—such comparisons work only between populations, not individuals, and then only sometimes. Time and again researchers have tried to find a way to pinpoint the particular individuals within a culture who will develop hypertension, and failed. No characteristic, no virtue or fault or combination, works for everyone. The vandal leaves little clues behind, pointing first to one suspect, then another. A grinning delinquent, high blood pressure, and clever—it backtracks, makes false trails, leaves no scent.

The direction all the evidence on hypertension is taking us is hard to accept; researchers glance over their shoulder at it, brush it in passing, but hate to put it in words: People who get hypertension are just different from those who don't. They have different metabolic characteristics, a physical roadmap unlike their normotensive cousins. This holds true only for some hypertensives—one researcher claims 40 percent—but for those it is vital. Insofar as the future of hypertension treatment goes, it's vital for the other 60 percent as well.

This group of hypertensive individuals are like nothing so much as genetic salt addicts. They are called "salt-sensitives," people who respond to sodium chloride in unfortunately pathological ways. ("Saltaholic" might be a more accurate word.) Salt-sensitives crave salt and eat more of it than people with normal blood pressure—and other people with high blood pressure. They have a lower "perception" of salty tastes, and so need to eat more salt in order to taste it in the first place. They also find the taste of salt less intense than a normotensive. Altogether salt-sensitive people eat between four and seven times more salt, given a free choice, than people with normal blood pressure. Salt-sensitives drink more water, too, and have a higher total sodium body weight. They gain weight faster, excrete less sodium, and excrete at slower rates.

It seems likely that the inherited trait involved in high blood pressure may be this invisible addiction to salt, coupled with a defect in the body's use of it. One European scientist calls hypertension a "continuous spectrum" of chemical sensitivities "which is not only inherited but also modified later by different levels of psychological stress or by different levels of behavior." Are salt-sensitive hypertensives suffering from a kind of allergy to salt? Do they have a higher set point for the metabolism of sodium? Eliminating salt from the diet can actually make high blood pressure worse (I'll elaborate in a moment). The most important question about salt in relation to blood pressure is the question of cause and effect. The kidneys of a hypertensive person need a higher osmotic pressure in order to function correctly—couldn't we assume that the hypertensive craves water and salt because those are the very things which increase pressure in the kidney?

What causes high blood pressure? No one knows. What cures, or at least suppresses, high blood pressure? No one knows. Certain things work for some people, not for others; the only clear-cut method of knowing what will work is to try it. It is an astonishingly widespread belief that salt causes high blood pressure and a low-salt diet relieves it; that once a person's blood pressure begins to climb they are doomed to salt-free soup for life. Clearly, sodium is involved, and the renin-angiotensin-aldosterone system is intricately related to sodium metabolism. The kidneys are involved. Blood volume is involved. But what happens, *really*, when a person with high blood pressure cuts back on sodium? With some people, their blood pressure *rises*.

The research is new, for no one had considered this possibility until a few years ago. There is a cautious consensus beginning to appear that only about 40 percent of the hypertensive population will have lower blood pressure on a low-salt diet. Thirty out of a hundred people will stay the same—a vain sacrifice—and the other 30 will have higher blood pressure. Martyrdom.

A person is probably salt-sensitive or salt-insensitive regardless of his or her blood pressure. Salt-sensitives may actually be lacking calcium. Calcium is another ion which moves about in the muscle cells, and has a profound effect on muscle contraction—and the blood vessels are muscles, too. Current research shows that some patients with high blood pressure can stay on relatively high-salt diets if they increase the amount of calcium in the diet to compensate.

Certain other people appear to benefit from potassium supplements and high-potassium diets. (Vegetarians are way ahead here; many fruits and vegetables are naturally high in potassium and low in sodium. Meat is high in sodium and benefits from salting.) In rats, restricted chloride with a normal sodium intake prevents high blood pressure from developing, apparently through affecting blood volume. The reason blacks tend to be hypertensive in greater numbers may not indicate a higher fraction of salt-sensitives, but simply a lower threshold of sensitivity—so that black sensitives, as it were, are simply *more* sensitive, and become symptomatic sooner.

Now look at the modern picture of treating high blood pressure. Most patients are immediately put on low-salt diets; in fact, the scare is so pervasive that lots of people with normal blood pressure put *themselves* on low-salt diets as a precaution. Calcium intake may or may not be encouraged.

When a person's salt intake is first reduced, his appetite for it is doubly heightened: "Salt is that which makes things taste bad when it isn't in them."

It's no coincidence that people who feel themselves to be in danger from someone might say, "He'll eat me up without salt." One requires a passionate appetite to hunger for salt-free food. Patients on a low-salt diet say nothing tastes salty and nothing tastes good. People asked to keep a record of their salt intake usually underestimate how much they eat by 30 percent. And still the world is a dreary place, meals a bore and frustration. Eventually, after as long as three months, a person's craving diminishes, and what was once bland tastes good again; what was once a desirable saltiness becomes unpalatable.

In fact, without the heady fear of high blood pressure and stroke, few people have a motivation to give up salt. A three-year Scandinavian study hoping to educate people into reducing their salt intake voluntarily met with little success. The only people who did change their habits were those who did not have, and had never had, high blood pressure; those people thought likely to benefit from low-salt diets made no changes in their habits at all, despite three years of education on the issue.

Potassium chloride, or KCl, has been virtually the only commercial salt substitute available for decades. It is not the best one. The best-tasting natural salt substitute is lithium chloride, related to the drug lithium carbonate given to manic depressives, a substance that occurs naturally in some water sources. (Accounts, some would say, for the different states of mind of folk from different places.) In the 1940s lithium chloride was given freely to patients to use in place of salt, and it was "received enthusiastically by physicians and patients alike because it satisfied the palate." The author of one report from 1950 sounds petulant, a tad irritable, when he comments on the Food and Drug Administration's decision to withdraw it from use in food. The blood lithium level of people using it as a salt substitute had risen to 2.9—0.9 to 1.5 is considered a therapeutic level for manic depressives—but only a few had shown signs of toxicity, and only one died. It wasn't entirely clear that the lithium was responsible for the death. The patients, anyway, enjoyed the experiment.

So potassium chloride is used, along with tricks like lemon juice and herbs. Not bad, except that potassium chloride tastes metallic and bitter—so bad that some patients needing potassium supplements prefer to drink pure potassium in liquid rather than sprinkle KCl on their food. And KCl used liberally, as it often is by salt-loving patients on low-salt diets, leads to hyperkalemia, or potassium overload, and serious, even lethal, heart arrhythmias.

There are possibilities on the horizon, brave new chemicals trudging proudly from the laboratories. One is an amino acid called glycinamide hydrochloride. It tastes salty—in some foods. It works best when mixed with MSG. "The fate of ingested glycinamide hydrochloride has not yet been determined," writes a researcher stiffly. And subjects say, reluctantly, it's nice, but it's not salt.

When aspartame, sold commercially as NutraSweet, was introduced, it signaled a new attitude. Aspartame is flipped sugar, a backhanded molecule that appears to pass through the body with little or no effect. Aspartame is a peptide—an amino acid combination—and Japanese researchers wondered what other synthesized peptides might be like. One group created 22 compounds with saltlike tastes. Two were not only as salty as sodium chloride, but *only* salty, without aftertastes or other flavors. Ornithyltaurine, one of the two, is composed of common metabolic substances. Research on ornithyltaurine will continue for some time. (The final vote on aspartame isn't in yet, for that matter.) It won't be on the market for a while, but the future of potassium chloride is finally uncertain.

Even the gourmet's mentor, M. F. K. Fisher, talked of salt substitutes. "There is, it seems, no substitute for NaCl. There is no faking its fine stimulus, its artful aid—except to use it with more respectful attention to its basic powers and dangers; except, perhaps, to taste it for a change, instead of taking it for granted."

If changes in the diet don't lower the blood pressure of the harried and hungry hypertensive patient, he or she will be given drugs—while staying on the diet. Some simply take diuretics, or water pills, to flush out fluid and sodium. (Unfortunately, the protector potassium is flushed out as well; diuretics have other side effects, too.) The next common category of drugs—and many hypertensive patients take two or more different medications—are beta-adrenergic inhibitors. The mechanism by which these drugs work is, like so much else, poorly understood. They reduce the force of the heartbeat, and thus the pressure in the arteries; they may affect the renin-angiotensin system. These drugs are hard to manage, requiring constant monitoring both to keep the blood levels high enough to work and to prevent an overdose. The right amount for any individual is often inside a very small range.

If such drugs don't work—that is, if the blood pressure doesn't drop—then vasodilators may be tried, drugs which relax and open the blood vessels so that the blood moves through with more ease. In tough cases, when the patient cuts back his favored salt and lives with nightly flushings in the bathroom, heart rate changes, nausea and decreased libido and the delight of potassium supplements, and blood pressure still soars, drugs which interfere directly with the nervous system are tried. Such medicines prevent normal nerve conduction, and force, ironically, relaxed muscle cells.

All in all, a picture of frustration and escalating danger, life lived with deprivation and diminishing returns—and for many, uncontrolled high blood pressure in spite of everything.

Two new categories of medication shed a slight light.

ACE drugs, or angiotensin-converting enzyme inhibitors, affect the production of aldosterone and therefore sodium handling. Again, they work for some and not for others and may be purely a drug for the salt-sensitive.

Calcium channel blockers are the second promising new category of medication. Calcium affects muscular contraction. Calcium channel blockers are chemicals that get in calcium's way, preventing it from crossing the cell membrane and participating in the electrical rush that stimulates the muscle's neural message. In some patients these drugs seem to relax the tense walls of the blood vessels, easing the force of passage.

A weird confusion arises with calcium, because it appears that simply increasing calcium in the diet—glasses of milk, yogurt and cheese with their sweet, salty flavors—has the same effect as the artificial calcium channel blockers, as though calcium had a built-in mechanism with which to recognize its own abundance. To top off the round of news, some people on calcium blockers who eat a low-salt diet get, for reasons utterly inexplicable, even higher blood pressure.

Simple questions merit simple answers. For a long time it was believed that the origin of high blood pressure was a simple question; pressure, after all, is a clear-cut proposition of physics, a straightforward matter of fluid flowing through a closed tube. But high blood pressure is an unfortunate matter not of physics, but of biology and desire; a matter of avidity and incontinence, and not only of cells and ionic potentials. It is an ocean of appetite and struggle, electrons vying for room, cells inflating and shrinking in the tides. From the deep, cold depths shivers a long and swollen furrow, rising, rising, to push everything in its path aside.

6

SWEAT

Wᴀᴛᴇʀ ᴀɴᴅ salt are so delicately balanced—they reach for each other and embrace; too much or too little of the one creates desire for the other.

Thirst is one of the most absorbing of experiences. It is a preoccupation in the truest sense, encompassing body and mind, tongue and lip and imagination, a physical mirage. Thirst is the inescapable and maddening image of water, coolness, wet, felt as dry heat. The body remembers water, its slippery cascade, the immediacy of water's first flush across shrunken cells. Hunger can have pleasure in it, be held at bay in sweet anticipation, whetted by smell and sight. But thirst is not so kind—the sight and smell of water actually hurt, until the thirsty thinks of nothing else.

"Thirst is the inner consciousness of the need to drink," wrote Brillat-Savarin. "When a man is thirsty (and as a hunter I have often been so), he distinctly feels that all the absorbent parts of his mouth, throat and stomach are involved in a parched craving. . . . The feeling is so strong that the word thirst is synonymous, in almost every language, with extreme covetousness and with imperious desire."

Desire, indeed: insatiable and constant thirst is a mythological punishment, a price paid for breaking a taboo, a madness-maker. "As cold waters to a thirsty soul," reads the proverb, and I can almost feel the parching of my tongue. Then there's

Tantalus, bound in hell near water that recedes from his touch. We have modern punishments as well. The poor manic-depressive, lithium bouncing through his veins like loose marbles, first too low to help and then so much in the blood his hands flap and tremble and he stumbles when he walks— lithium makes him piss buckets full of urine, leaving him thirsty all the time.

In the hypothalamus in the middle of the brain is a small place called the thirst center. It responds primarily to two changes in the body: shrinking cells and increased osmolality. Remember osmosis, and the motion of water to balance concentrations. Osmolality rising in the body is analogous to a glass of water filled with marbles—the marbles representing the blood solutes, like sodium and minerals and proteins. Add marbles, or remove some water, and the result is the same— proportionately more marbles to volume of water. Certain body cells are designed to detect such changes, and they in turn stimulate the thirst center. With all our liters of salty, bloody water, we can tolerate only about 1½ lost cups of liquid before we're thirsty. All at once I wander to the kitchen, barely considering, and reach for the tap. (I don't consider the luck of having a tap in the first place, of not having to spend half of every day treading to the river and back, licking my lips as I pass the forbidden irrigation pipes.)

The body gives up its precious water and salts in urine and feces and sweat and the exhalation of air. For reasons unknown, the body knows its needs. Dry mouth is a peculiar symptom, local and immediate and almost irrelevant to the gasping internal cells. This local sensation has no local relief, either: merely washing the mouth out is not only worthless but a painful tease. Thirst is satisfied by water drunk, and nothing else, and in fact the sensation of thirst disappears when the proper amount of water is swallowed, before the shrunken cells expand. The phenomenon is called "preabsorptive satiety." Before the molecules travel to the tissue, when my stomach is still sloshing and full of the fresh drink, the thirst center is shut off, with a silent sigh. Recognition, perhaps; the water waving its yearned-for hydrogen bonds all the way down my gulping throat.

IN A HEALTHY person, thirst and sweat are entangled; unlike thirst, sweat often remains silent. We sweat three ways. There are in our every moment tiny droplets of water evaporating from our skin and lungs. This is called with unconscious irony an "insensible" loss. In conditions of heat and exertion we produce thermal sweat from specific glands—up to 12 liters a day, a few of us, and as much as four liters in an hour in the short run. And there is another form, sometimes called palmasole sweat for the areas of concentration—a sweat of emotion and excitement, fear and love. (It has been elicited in research simply by assigning volunteers problems in arithmetic.)

(A few unlucky folk suffer hyperhidrosis, or constant and profuse sweating, usually from the hands, feet, and face. Like so many of the worst problems that don't kill but perhaps should, it affects teenagers the most. Victims are instructed to coat their skin with tanning agents, as you would leather and hides.)

Sweat is a source of considerable interest, but it poses problems. It is easy enough to make, but how to collect it? An experiment requires precision and reliability. A *good* experiment must be replicable, consistent, dependable. Experimenters have tried a number of methods for catching sweat off a damp body: plastic collection bags taped to the underarms, a distilled water wash, filter paper pasted to the skin. One experiment involved dressing subjects in long underwear, having them exercise until the underwear was soaked in sweat, then washing both the subject *and* the underwear with a measured amount of fresh water. Others have resorted to inserting tiny tubes directly into individual sweat glands, like mosquito noses.

(Writes one researcher, "None of these methods is totally acceptable.")

The human system of sweating is unique. While some other mammals have eccrine sweat glands like ours, none have eccrine glands in such quantity or use them for heat dissipation. Whether or not the human method is superior is controversial; some researchers claim it to be a marvel of inefficiency. The skin glands of mammals fall into three categories: sebaceous glands secrete a protective layer of oil to keep the skin moist, apocrine glands release scent signals, and eccrine glands release moisture to keep the skin soft, sensitive, and able to grip. Primates have eccrine and apocrine glands scattered together across the surface of the body, with the eccrines clustered on the hairless surfaces, like the palms. There the damp excretion aids in gripping, to keep a swinging monkey from sliding down a vine. This is true for primates, that is, with the exception of humans. Our apocrines have virtually disappeared, leaving us curiously dull animals when it comes to odor—though the tide-swept fetus begins to develop apocrines early on, only to abandon them. (As for smell, it is human emotion that stinks: in research on new deodorants, subjects are sometimes stimulated into anger and upset by psychologists, in order to produce the necessary bouquet.) Instead of apocrines, we have enormous quantities of eccrine glands, several million of them—more than hair follicles—and they are almost everywhere. The only sweatless areas on a human are the ear drums, the nailbeds, a part of the lips, and the bulbous end of the penis. On the soles of the feet and the palms of the hands there are more than 3,000 eccrine glands in every square inch of nervous, clammy skin.

The sweat gland has often been compared to the kidney, a similarity only partly metaphorical. Eccrines are, in a sense, quite simple, consisting of a large clear cell, a structure called the secretory coil, and a tubular duct. The cell and coil secrete an isotonic, or neutral fluid no more or less salty than plasma. The tubular duct, on the other hand, varies the secretion and reabsorption of sodium chloride, affected in part by the ubiquitous aldosterone.

Sweat is nearly all water, but is spiced with sodium and potassium, chloride, sugars, urea, and a few other esoteric wastes. The concentration varies from person to person, and depends on the rate the sweat is produced—which depends on heat and effort. Thermal sweating and the cool evaporation that follows is the only way the human body can lose heat when the air is hotter than the body itself. It is not chance that our temperatures are so high; we're not very efficient above 98.6°. In humid heat much of the effectiveness of sweating is lost; the skin breathes wet air like the membranous unborn, but in a suffocation.

Such a heady vapor is considered pure by some: the Klickitats, the Klamaths, the Sioux, Crows, Arapahoes, the Yakimas and Takelmas and other tribes made sweat lodges. They were vital structures made from a careful plan, the door always facing east. Some lodges were made of willow, others of planks or dug partly underground, and constructed to block out all light. The floor of a proper sweat lodge is covered with pine or fir boughs, and heated rocks are piled on one side. In the sweat lodge medicine was worked, prayers said, luck invoked, and health restored. Sorrow was poured out with the sweat, impurities were flushed from the pores and the soul, the hunt was made fruitful.

It is devilishly hot in the lodge, in blackness, with cold water splashing on hot rocks into sudden gushes of steam, dried leaves of sage burning, the grunts of the men shifting their damp weight. A white man invited in one winter night called the darkness that of ignorance, the earth's womb, a terrifying place: "The noise of the steam, the heat, the lack of air, the bitter taste of sage in the mouth, and the sting of sweat in the eyes, all in pitch darkness . . ." until the plunge outside through the fur-skin door, into snow and cold and moonlight and the river running nearby. Then, he wrote, he felt "like an angel, with no bones."

CHILDREN ARE CLOSER to their mammalian origins than adults, and it shows in how they sleep. In cold weather children curl into balls, their small bottoms up in the air to meet the coolness with an insulating layer of fat, heads hidden under arms and their imaginary tails wrapped round themselves. In heat they spread themselves out, arms flung and legs hung loose, wet and evaporating, offering their skin's sweaty surface area to the stirring breezes.

One liter of sweat evaporated removes over 500 calories of heat, and as much as six pounds of weight. It's not a perfect method—the ill-tempered camel is far more advanced in its biological tricks. But it works. Birds, lacking sweat glands, dip and bathe to cool themselves. If the sweatless elephant can't find water to spray on himself in the sun, he'll use his own spit instead. And consider the dry-skinned dog, who must pant to lose his heat. Panting can raise the breathing rate to more than 300 times a minute.

Why is the human so different in this particular respect? Eccrine glands aren't really that useful. Water and minerals are lost rapidly in sweating in the conditions most requiring their conservation: heat and activity. Elaine Morgan hazards an evolutionary guess: eccrine glands are an old development, the result of a marine period in human evolution, a protection of the organism against a harsh salt environment. The sweat gland did not, perhaps, evolve as a method for dissipating heat, but as a method for dissipating salt.

The sodium content of sweat varies a bit—from 18 to 97 milliequivalents in a liter. (In cystic fibrosis sweat sodium typically ranges above 60.) In periods of heavy work or in unusually hot environments, the amount of sodium first rises, then drops, but the more profuse the sweating, the higher the salt loss. The rate of sweating doesn't vary with the body's available water—dehydration alone, the stimulation of the thirst center, won't stop the loss of water in sweat. You sweat in order to lose heat, and the hotter you get the more you sweat.

Men working in warm environments (or simply resting in

hot ones) may lose between one and six liters of water and as much as 10 to 30 grams of sodium in a day. A loss of 10 percent of the body's water can make a man delirious and ill. A loss of a little salt, a few grams an hour, can too. (Forgive the male reference; much of the research on salt and water needs in hot environments has been done on marines, called "voluntary subjects" in the literature. They lift weights in steam rooms, dig ditches in the tropics, run in the Arctic, and lie, like mad dogs, in the noonday sun.) To prevent the devastating changes in osmolality, it is best to balance salt and water intake; if one is limited the other is best limited as well. Either way, the body does best with preparation for exercise and heat, by "loading up" with both salt and water beforehand. The sweat gland itself is deformed with heat, shrinking as the stores of glycogen, a fuel sugar, are burned up. Though the shrinking reverses afterward, when fuel is restored to the body, salt loading seems to slow the process of shrinking in the first place.

Whether or not sweat is an efficient evolution, humans often seek the experience of it. The lipid shine, the oily smoothness of hot sweaty skin and the lean tired feeling of achieving it; and the smell of it, too, the pheromone-laden air of a room damp with sweat and effort; turning to a fan for the wind across a wet chest. Sweat is sheen, color, fatigue, and motion, the pleasant worn sensation of stretched muscles and the deep suck of oxygen. Sweat is sticky and slippery at once, a layer of the elemental when two bodies meet.

ATHLETES IN competition sometimes grow feverish with their toil, with temperatures as high as 106° F. recorded after races and contests. (Most people function poorly with a fever over 102°.) Football and basketball players may lose between three and seven pounds of body weight in a single game. They are familiar with heat cramps, the exquisitely painful spasms of muscle that comes with salt loss, attacking the back, the abdomen, and the limbs. Heat cramps last hours and cause agony.

They are so reliably induced by exercise in the heat that certain types of workers are especially susceptible: sugarcane cutters, boiler room firemen, coal and diamond miners, steelworkers. (And volunteering marines.)

Now we must give a nod to the lowly camel, the glorious and amazing camel. (Only the two-humped, by the way, should be called a camel. The one-hump is a dromedary. But I'm about to generalize, and beg forgiveness.) A camel's hump is a store of fat, a perfect horn of plenty. Each hump can weigh as much as 30 pounds when replete. Camels can go more than two weeks without a drop of water, and lose 30 percent of their weight in the meantime. The humps gently shrink and the fat is burned, releasing hydrogen. And the hydrogen combines with the oxygen from the lungs and makes water, fresh and pure. Camels have a body temperature often lower than the air, and when resting in the sun will lie in one place all day, thin legs tucked neatly up beneath, facing the vertical rays with their round bodies so that the sunlight deflects off them. They are born to physics, camels, and do odd and unexpected things. A group of camels will lie in a tight cluster, side to side to shade each other, and move only to follow the changing angle of the sun.

Camels humidify their inhaled air to keep it cool, and before exhaling cool it again, recovering the tiny droplets of condensed water in their nasal passages. They urinate infrequently, converting some of the waste urea back into amino acids for nutrition. The urine, when it finally falls, may be twice as salty as the ocean. (I have heard the rumor, unverified, that camel urine kills head lice, and makes a good shampoo.) Camel feces are dry and powdery, milked of excess. Camels neither pant nor sweat except in the most extreme heat: a camel's temperature rises 16° F. before sweating begins. A man would be long dead and dry.

When a camel emerges from a trek, shrunken and thin, he goes to drink. In 10 minutes he will easily drink 30 gallons of water—his stomach can hold twice that. His red blood cells expand to 250 times normal size like sponges, soggy and moist. This flood is distributed evenly, packed away, in less than 48 hours. The hump fills fat and sassy in a day or two, and the camel spits in your eye and looks off to the distance, disdainful of all things human.

But humans can change, in wee bits. It is for the best we have big brains, because our bodies are good for so little. With exposure to heat every day for a spell, in a week or two the body sweats more, distributes the sweat more evenly, and the sweat has less salt. The person's subjective discomfort diminishes, too; the heat feels less hot. Heat cramps can be avoided with the right combination of water and salt. Too much water and you can grow waterlogged like the aquarium-sucking schizophrenic. Take salt tablets alone and you risk stomach cramps as the water rushes in to dissolve them (and the sudden salt load leaves you thirsty). The simplest, and probably the best, method for preventing salt imbalance in heat is to dissolve a salt tablet in a glass of water and drink it down. Sugary drinks like Gatorade, loaded with free electrolytes, are merely an expensive version of salty water—or the equivalent of beer and pretzels. For real acclimation, study the professionals: the bedouins.

The water in the wells at Tarfaya in the Sahara is salty, ranging as high as 12 grams of salt in a liter. The bedouins who pass by in a nomadic cycle (the ones not yet brought into the urban fold) drink curdled camel's milk most of the year, relying on the deep well water only when the milk supply diminishes in drought. Even camel's milk is drunk only with meals, and altogether a man's daily intake of water may be little more than half a liter when a tribe is stationary. Water and salt conservation run strong in their traditions. In the Muslim time of Ramadan no liquid can be drunk from sunrise to sunset, not even one's own spit. If a man crosses a dry stretch of desert and arrives thirsty at a well before his friends, he waits without water until they come, lightly resting in the

dry control of salt's residue, and then everyone drinks at once, with ceremony.

Claude Paque, a French anthropologist, studied the bedouins in the 1950s—their habits, their traditions, even the amount of sodium excreted in their urine. The bedouin nomads eat little free salt, seeming to have no appetite for it, complaining it gives them headaches. "I only take salt when I need it," a woman told Paque, comfortable in a tent with an interior temperature around 40°C. (94°F.). The bedouins sweat little, long accustomed to salt and water concentration, with nearly saltless urine. Paque wrote of the first time he drank, with trepidation, from a salty well: "It seemed to me that this water, cool and salted, quenched thirst more rapidly and more definitively than pure water. Under similar circumstances, I was accustomed to drinking pure water for hours to quench my thirst completely. During this trip I continued to drink salted water. It seemed to me that I was then less thirsty, and sweated less."

An engineer named Daniel Hillel helped bring irrigation water to a Jewish settlement in the Negev desert some years ago. He describes the first celebration, when sprinklers were turned on. Bedouin noblemen from the area had been invited to the ceremony, and stood watching the long-absent Jordan River water spray into the bright sunny sky.

"An old Bedouin sheikh, befuddled at all the strange gadgetry, deaf to all the highblown speeches, was suddenly caught in the swirl of water," writes Hillel. He paints a picture of the old man, soaking wet, with robes hanging, his beard dripping—perhaps for the first time—with water. "Finally the hapless old sheikh mustered a resigned and kindly smile as he blurted out: 'W'Allah el Azim, hada'l mattar min el ard!'" Great God, this rain is coming up from the earth!

7
GEOLOGY

Gᴏᴅ ꜱᴘʟɪᴛ ᴛʜᴇ waters, raised the firmament, struck the sky with lightning, wiped his damp brow, and spilled sweat to make the people. Not a bad story. But where did the salt come from?

Sodium chloride is called, when it appears as a mineral in the earth, halite. The linguistic origins of the word halite are varied: *hals* is the Greek for salt, and that's obvious enough. But what about the possibilities raised by similar roots? The Middle English *hali*, which means holy, or the Latin *halitus*, for breath. There is the Latin *hallucinari*—"to wander mentally"—which is probably stretching it a bit. The Greeks offer several more: *halos*, a ring of light, rising in its turn from *alos*, for halo. There is *halten*, to hold, and *halma*, to jump. But given all the choices, my favorite is from the Hebrew: *hallelu*. Praise.

So if people are born from sweat, and pigs from urine, then what about salt? Dried tears, perhaps, the ones that spilled too far to be part of the salty sea. Mountains of salty tears turned to white gold, bright and hard and solid as ice.

Aʀɪᴄʜ Rᴜꜱꜱɪᴀɴ merchant had three sons; two smart and one, Ivan, foolish. When they were grown he sent the two clever brothers to sea with good ships and stalwart crews, to find their fortunes. He looked at silly Ivan and shook his head.

"For you, boy, a ship and crew as well," said the merchant, and pointed to the harbor, where a decrepit old ship floated, rocking its ill-kempt sailors. "Find your fortune, boy," he added, and turned on his heel.

Ivan was too foolish to be angry—too foolish, in fact, to notice the condition of his ship and men. It was a piece of Ivan's foolishness that he enjoyed life, and suffered others gladly. He immediately set sail for parts unknown, happy and sure of adventure.

Before long a storm blew Ivan's ship out of the byways of the ocean lanes. (He wasn't much good at navigation, it seems.) After a long wet night in the waves, his ship struck land.

Ivan climbed out of his boat into a land of amazement, an island that was no more than a great mountain, shining so white in the sun it hurt the eyes. The mountain was no more than salt, but what salt! Pure white, flowing like sand off the fingers, full of sweet and pungent flavor. And there was enough to feed the world. Ivan knew a good thing when he saw it, and ordered his men to empty the ship to its timbers, fill every nook with salt, till the ship could hold no more.

They set sail again, to find a market for their salt, and landed on another uncharted island. Ivan proudly carried a basket full of fine white salt to set before the island's ruler. But the ruler had never seen salt before, and laughed in the poor boy's face. "White sand! This is absurd!" he hollered, and kicked Ivan out.

That night Ivan slipped into the kitchen when the cooks were busy in the pantry. Pot by pot, dish by dish, he sprinkled the lovely salt in the king's food, and scurried out. The salt dissolved and grew invisible, like all good magic.

Ivan waited in the crowd outside the castle while the king ate, a grin on his silly face. Ivan had no doubt of the salt's power to convince. Sometimes, thought Ivan, you just have to *show* people.

The king suddenly appeared on the balcony, all excitement.

"My cook has outdone herself!" he shouted. "Such food I've never tasted before! Such delight and savor! Cheer your royal chef!" But the honest cook was as bewildered by the taste of her cooking as the king, and shook her head.

"I didn't do nothing, Your Highness," she quibbled. "It's the same as before. Someone else must have done it."

And at that Ivan stepped forward with his leather pouch of salt, and the rest hardly bears telling. Ivan exchanged his ship of salt for a ship of jewelry and gold; he grinned his foolish sweet smile at the king's daughter, and she fell in love with him; and together they sailed back to Russia, where Ivan met his father and brothers again, and was kind to them, as usual.

SALT IN THE earth is an odd bird, so solid in appearance, so unyielding—but plastic and alive in its behavior. From its origins it was lively, born when the ocean floor itself was frisky, yawning and rolling and grinding out its ribbons of rock. Great undersea mountains rose and cut off the flow of currents. The early world was hot, very hot, and in shallow places the water quickly evaporated, leaving a salty brine. Sediment fell and the salt was buried, pressed tightly and deeply under the rock. This process continues even now, in the subtropics, where rainfall—rich as it is—can't keep pace with evaporation. Gradually (in 100 million years and more) the rock salt is buried under fresh sediment, deeper and deeper, and sometimes buried still further under the shifting borders of the ocean. Salt, like all sediment, is buried in repeating bundles that reflect the geologic eras—and how casually we discuss those eras. "One hundred million years"—an infinity. Macallum said of the paleoratios that "all the serried ages of the earth's history do not sleep in stone alone," and salt is one example: the ages of the earth sleeping, being put to sleep, at intervals. One of the biggest continental rifts in the world's birth occurred about 230 million years ago, and in that one brief spell about 10 million cubic kilometers of salt were formed, and slowly hidden, most of it still resting along the sea's shelf. Ten million cubes of salt, each a kilometer along the side.

Salt rises. The exact reasons for this are complicated and a little strange, so bear with me. The fact is that these enormous fingerlings, called diapirs, work their way to the earth's surface, where—in the right conditions—they appear. (The dome beneath the emerging diapir has long been called the "salt mother.") Under thin layers of topsoil diapirs look like dirty flower blossoms, deliberate, patient, beautiful. While diapirs are still under the ocean they sponge up oil and gas (about four-fifths of the United States oil and natural gas reserves are associated with salt deposits) and store it as a cactus stores water.

These gargantuan pilasters, these ridged and vertebral columns, are miles across and miles deep. Most rock salt is halite, sodium chloride. The great salt domes and beds are almost pure halite, deposited in layers. It is rarely Precambrian and in fact is almost systematically positioned in sediment, reflecting as it does the methodical shifts of the crust and the flood-and-drought cycle of the atmosphere. It is timed into the earth's crust like marbling in a cake, and slips in everywhere: Kansas, Oklahoma, Texas, Michigan, Pennsylvania, New York—and Europe, Iran, the Sahara—lie unsuspecting atop massive beds of salt in motion, slowly. Some domes are beyond measurement; their depths can't be sounded.

The first published account of a salt dome was in 1856. A French mining engineer in Algeria, a man named Ville, found first one dome and then several more. He noted, with a geologist's sense of history's swiftness, "eruptive and geyserlike behavior." Salt mines were hundreds of years old, Europe strode across dozens of major domes, but until Ville, no one had imagined their size, their purity, or their vitality.

Arrhenius, a Swedish scientist, proposed a theory: that salt rose and moved because it was buoyant, like a cork floating on a sea of rock. Other theories, as odd and oblique, followed: salt was pushed out of the earth by faults, salt increased in volume as it crystallized, salt was thicker than the surrounding material. The truth is stranger still, but Arrhenius was closest.

Salt is almost incompressible—it is as big as it is big, and in this sense truly unyielding. Struck with a sudden shock—a hammer, or an earthquake—solid salt can shatter into fractured pieces. With the right kind of blow salt can rebound, too, spongy and elastic with the force. But under a constant strain, not sudden, not sharp, but continual, salt becomes a strange and lovely liquid, mutable, amoebic. Salt in the earth in all its enormity flows like lazy water upward, rising like cream in milk and oil in vinegar. It rises in multiform columns through the heavier, denser sediment, a forest of gigantic salty trunks. A crystalline orchard of rock.

There are three kinds of rocks in this world, as you and I learned in grammar school: igneous, sedimentary, and metamorphic. Igneous rocks are the product of crystallized molten material; they are mostly buried, and constitute about 95 percent of the earth's rocks. This is basalt, those impossible rows of columns sheared off mountainsides, and black glassy obsidian, which forms whole mountains of its own, and quartz and pumice and a few others, like certain granites. The sedimentary rocks, like halite, are the compressed remains of particles—what one author calls "the refuse of all earth surface activities." The geologist's prose really starts to fly with the sedimentaries: limestone and graywackle and chert, tufa, gypsum, marl, breccia, travertine. Sedimentary rocks are, for the most part, exposed to the air and soaked, like salt in oil and water. They are marked with fossils, footprints, even the tiny depressions of ancient raindrops. The relatively few metamorphics are the mutations of the other two, the result of pressure, temperature, chemical reactions, gorgeous teratogens creating marble, slate, hornfels, gneiss, and more.

To study the sedimentary salt is to study sediment itself, and in this case that is mostly sodium, the little ion that just keeps wheeling through, in and out of the ocean, in and out of rock. There are 120 trillion tons of it in the earth, spread out molecule by molecule in the dirt and shore. Changing coastlines have left salt pans and salt flats and salt lakes, shallow and dusty white. Even the ice shelf of Antarctica is seasoned, holding a plain of frozen salt, hundreds of tons of fractured crystalline cubes. People cut blocks of salt like ice for market; you could build an igloo from them for the desert, a nomad structure till the first rain. Sodium saturates soil as salt saturates water. When an animal dies and begins to decay into the ground, the sodium in its bones leaches out until the bone and earth are in a steady state, equal. An archaeological chemist can plot fossil sodium backward to find the fossil's age.

Now, to understand salt's motion is to learn a little about crystals. A crystal is anything made of atoms arranged in a repetitive pattern, and that is almost every solid around (with the exception of glass and plastic). Imagine atoms as spheres (spheres of influence, with wavelike orbits on the surface) and visualize crystals as repeating stacks of spheres arranged in a particular order, a different order for each kind of crystal. Water is water *because* it is water, because in the tautology of chemical structure water is hydrogen bound in pairs to an oxygen, and all pairs of hydrogen bound to a single oxygen make water. So are crystals. (So are we.) The qualities of the crystal—shape, symmetry, electrical conductivity, ability to be compressed or to cleave in flakes—depend on the material of which it is composed and the order in which that material is placed. Crystals are products of repetition, of pattern, harmony, method; they are prisms of order and alignment.

Salt is sodium and chloride, packed round each other in a planar network so that any ion of sodium is surrounded on six sides by an ion of chloride. Think of it as front, back, both sides, head, and foot. Wherever the sodium reaches out in its sodiumness, it encounters chloride in all its chlorideness; they bind in ionic bonds. This gives salt a four-fold axis: it has four 90-degree angles. In contrast to salt's simple elegance, molecules of diamond and graphite and silicon dioxide have complex, angular structures, still repetitive but forming chevrons, wedges, and dovetails.

Because of their repeating nature, crystals diffract—bend—waves (at least in theory; practice, on the atomic level, is more problematic). Salt was the crystal used early this century by the father-son team of scientists, Sir Lawrence Bragg and Sir William Bragg, to determine crystal structure using X rays. Crystals can conduct electricity—thus, the crystal tube in radios.

The arrangement of sodium chloride, such that every ion is surrounded by its antipode, is one of the most obvious and direct responses to the problem of electrical neutrality. There are literally no *molecules* of sodium chloride in salt. Given a tool small enough to slice off a pair, an ion of sodium bound to an ion of chlorine, you couldn't cut out a molecule of salt, of *NaCl*. Salt is a continual pattern of ions in relation, of—seen in its necessary three dimensions—NaClClClClClCl, or NaNaNaNaNaNaCl, or better yet, NaClNaClNaClNaCl, because in salt every sodium is surrounded by chloride, every chloride surrounded by sodium. Cut out a "molecule," a piece, and you no longer have that self-contained blanket of opposition. Salt is its own *environment*, a crystalline atmosphere, a simple crochet of twins.

At 150° C. salt rebounds like a sponge cake; at twice that temperature it begins to flow, and the sodium and chloride atoms stacked atop each other like bricks start to slide, slip, slip, like hands rubbing together, like molasses running through a sugar cube. An ion of sodium is only half the size of an ion of chloride, so they slip across each other like wheels over cobblestones, bumpy. Temperatures rise quickly in the unfathomed depths, miles and miles down into the center of the earth, and rock salt is both a conductor of heat and less dense than the surrounding sediment. The conditions are ripe for motion; it rolls and sallies and drips upward, syrup between molecules of rock this and rock that, edges in like ocean breakers nudging under driftwood, slippery, viscous, flowing.

(Granite, of all things, behaves in similar ways. It is less plastic, but slides plane to plane like palms rubbed in anticipation. Unconcerned and steady, granite rises; it is filled with water and gas, and when granite reaches the air its emergence is violent. Lava and volcanoes are not uncommon. A salt current seems placid in comparison.)

The penchant of salt domes to hold oil and gas like earthenware vessels of wine struck certain scientists as a sign from nature: What better place to store nuclear waste? There are hundreds of salt domes along the Gulf Coast alone, and all that's needed to hollow them out is the simple pump and pull of brine mining. The cost of such storage would be a fraction of the expense of above-ground methods. Morton Thiokol, its roots in salt, for several years has leaned on its production of nuclear weapons for a big share of profits, and looked from the resulting waste back to its own salt deposits for ideas. When it rains, it pours.

Too bad the salt won't cooperate. Part of the problem is the motion, the restless lift and tuck of a dome at work. The rest is in the nature of salt itself. A salt dome in Texas used to store natural gas developed a leak and an entire town had to be evacuated. A giant sinkhole in another part of Texas that began as a small crack turned into a cavern the size of four football fields in six months, all the result of groundwater dissolving through a dome. Several years ago an oil drill pierced a cavern of salt in Louisiana and half a lake disappeared down its drain.

More ominously, a report by two geologists reveals the evanescence of the salt, and thus of domes. The scientists radiated salt with the levels of gamma radiation and heat expected from nuclear waste, and found that they evoked atomic charge changes in the salt's ionic structure. Highly alkaline and highly acidic brines were both created, corrosive enough to eat through the storage casings. In an echo of Ville's comment on salt's motion, one geologist commented, "The salt is extremely reactive." It seems that salt won't hold rock-still when it should, won't move when movement's wanted. Pulled by a thread, it won't unravel.

THE DOMES are so large, so deep, and their generation so unbearably slow that understanding the forces at work is necessarily part guesswork. Physicists and geologists have used laboratory models of varying sophistication, and now use computers. I have a photograph of an invented deposit, made of silicone putty and a transparent plastic, mixed and spun in a centrifuge in an imitation of gravity and time. The result is fine and intricate, a splendid series of troughs and valleys walled off from each other by thin, cellular walls, laced with distorted concentric circles. Such lines are revealed in old desert diapirs and the roofs of salt mines, lines of circulation within a rising dome, lines of encapsulated sedimentary rock sucked into the salt's center and captured.

Two English physicists, Rayleigh and Taylor, developed a theory of salt-dome formation. In their view, a deposit begins (and such things are always beginning, buried in the earth) with two flat layers, heavy on light, gradually undulating into a series of small mounds. The mounds are spaced according to a "wave length" dependent on the variations of density and structure in the overlying layer. The slower mounds give way to the faster, which grow into hillocks, and then cone-shaped hills, dramatic. Their rise is periodic, separated by times of rest. As the light salt rises it reaches less compacted, looser sediment, and the pressure of buoyancy changes, till the cones flatten into bulbs, the bulbs into pancakes that stretch out beneath the surface in canopies that sometimes merge. Some domes form giant thumbs, others bulbous balloons and peaked mushrooms, depending on the strength and thickness and weight of the material in which they rise. The diapirs cluster in polygons and clumps, hidden art. If you could sculpt it out, cut away the rock and leave the diapir shapes alone (*diapir* is Greek, to pierce) you would have a soft and female thing, voluptuous, curved like wood turned in a lathe, layered and molded the way clay in a child's hand is shaped by the rings of its fingers.

Don't you know you sit on salt? It hides, we ride it. Only in deserts are the tops of diapirs visible; in more temperate climates groundwater dissolves the ones that reach the surface, in a process called decapitation. They rise for 15 million years, they rise for 50 million years, and slide beneath soil. In the Kavir desert in Iran is a beautiful diapir head, a deformed circle of twelve domes crowded against each other, 50 million years old and 40 kilometers across. Its edges are twisting zebra-stripes of shale and gypsum and salt.

Sometimes salt spills out, pouring down mountains like the Zagros in the Persian Gulf, variegate chrysanthemums scattered among the peaks. They are glaciers, their salt is more than 500 million years old, and some rise more than a kilometer high. These glaciers move quickly, sliding down the mountainsides a meter or two every year, faster than diapirs for two reasons. One is brute force, the tectonic crash of two continents in collision. One is light and chemical: the Zagros' salt is so old, has seen so much in its time, that the crystals are greatly distorted by odd pressures and friction, so rough and tumbledown they resist the flow crystal by crystal, the way a lumpy rock won't roll. Now and then it rains a little in the Zagros, and the water dissolves the surface layer of old crystals, which dry and unite into new, malleable shapes, amenable to motion, glad to be on the move. The glaciers, enfolded like drawn curtains—like "pushing a napkin through a ring," writes Christopher Talbot, who puts markers in the salt and watches them move—start and stop with the weather. Fits and starts, a geological hopscotch game. "A purely aesthetic appreciation of salt formation is hard to avoid," he writes, almost grudgingly. I can see him now, hard-booted in the rough rocky soil of Iran, his eye on a jumpy flag traveling down a mountainside like lava in a freeze frame.

8
SPRINGS

"A<small>S IS FABLED</small>, there is a lake in Palestine, such that if you bind a man or beast and throw it in it floats and does not sink. . . . They say this lake is so bitter and salt that no fish live in it." Thus wrote Aristotle, telling tales again, and, as usual, right. The Salt Sea, the Dead Sea, the Sea of Araba, the Sea of Lot, or Mare Asphaltum, a Roman name given because in the Sea floated chunks of asphalt, harvested and sold to the Egyptians for their secret and arcane rites of embalming. The Dead Sea is the lowest body of water on earth and as salty as water can be—26 percent salt by weight. It is the remnant of an ancient lake, the Lisan, and is full of clay and chalk, banded by marl and gypsum, surrounded by a valley with Pleistocene deposits 3,500 feet thick. Looking over the Dead Sea is Mt. Sodom, a salt dome 220 meters high, webbed with chimneys and caves and grottoes sketched out by rain. The twisting, skinny, thirsty Jordan River waters the Dead Sea, which has, like the Great Salt Lake in Utah, no outlet; evaporation keeps its size steady. Except at one shallow end, the Dead Sea is 1,000 feet deep, eerie in its emptiness all the way down. There is more than sodium and chloride dissolved here: the acrid, burning waters hold bromine and sulfate, potassium, calcium, magnesium, carbonate, and silicate, and the water hurts the eyes, scrapes the skin like wet pumice in a liquid form. Along its shallow edge circular bergs of salt appear, rime and crystal, a field of white mushrooms growing in blue dead water.

The lake evaporates continually in the hot dry air, and sometimes the sky is hazy with rising salt vapor; the sea's elevation—1,300 feet below sea level—means there is increased atmospheric pressure in its valley, and extra oxygen. Some people believe the bromine in the water makes a natural sedative, and the difference in atmosphere makes for safe sunbathing. The waters are said to ease rheumatism, and the hot, oxygen-rich air is believed a palliative for psoriasis. The coast of the Dead Sea has health spas, hot springs, luxury hotels scattered along it, service to the tired-jointed tourists who come to bathe.

(Yes, she's there, the impulsive wife who turned for a backward glance; pointed out variously as this pillar or that column or that salty cervix rising from the sea banks like a tiny dome, deckle-edged and lumpy.)

SALT SPRINGS have always seemed a gift, and unexpected. The bubbling water was called the "mother liquor"; it gave birth to salt. Aristotle tells another story, that Hercules came to Chaonia and offered its inhabitants fish. But they chose salt instead, knowing a good deal, and a spring rose, which still in Aristotle's time yielded salt "not in lumps but loose and light like snow." In Germany a spring was discovered by a pig— how is not explained, perhaps it sniffed the fragrant waters out—and the locals were so grateful they built the pig a special place. Dead and stuffed, the pig rests in a glass case, immortalized by a plaque describing its achievement, for which it "acquired for itself imperishable glory." The Chinese used geomancers to divine for springs; Americans drilled, wherever they traveled—or, more wisely, asked the Indians. Several times men looking for salt springs found oil, and were disgusted with the useless black syrup. In the nineteenth century, when salt became a competitive industry, a man drilling for oil in Kansas found one of the great solid domes; he changed his plans, and opened a mine. Once in West Virginia a salt spring came through the drill in a geyser, carbonated with flaming natural gas; when the fire was extinguished and the brine and gas separated, the gas fueled the furnace that boiled the water until it evaporated and left behind its salt.

The best springs are the thermal springs that dissolve the edges of a dome before leaping to the surface; they're pure, uncontaminated by other minerals. (Pure for salt, that is; not so popular with bottlers of mineral waters.) The methods for getting the salt from the water haven't changed much, in spite of machinery. A portion of the *De Ra Metallica*, a medieval description of metallurgy, said this: "Some people make salt by another method, from salt water which flows from hot springs that issue boiling from the earth. Since the water flows continuously from the pool through the little canals, and the spring always provides a new and copious supply, always boiling hot, it condenses the thickened water poured into the pans into salt." So it went, and goes.

In the late 1700s the Onondaga Indians in New York sold a salt spring to the state, with a contract guaranteeing them a share every year, measured in the tons. By 1801 the spring was producing 62,000 bushels of salt each year and the surrounding town was called Syracuse. Salt is hard to transport at a profit, heavy and cheap at the same time, and the Erie Canal was built to carry the salt barges from Syracuse to the coastal cities.

Onondaga salt was a low-quality product, but with little competition and its own canal, it sold well. Gradually the wood furnaces that stoked the boiling pans deforested the entire area. (A salt spring in Michigan deforested its surroundings, too, for fuel and cooperage stock. When the supply of trees was exhausted the salt industry turned to cotton and burlap bags. These slowed production, and raised the price.) Partly in response to the need for wood, a solar evaporation project began in Syracuse. Solar salt was kept duty-free to encourage its production. By the middle of the 1800s Onondaga produced over three million bushels of salt by boiling, and more than

200,000 by evaporation. But competition increased, especially foreign salt from the West Indies and England. Smaller New York springs went into production. The growing Chicago meat industry, too far to be worth the cost of transportation from Syracuse, gave a big boost to Midwestern salt producers. Onondaga's salt had never been the best available, with a slight reddish cast and a tendency to cake. Efforts at purification helped, but added expense. For all these reasons, Onondaga declined; the Erie Canal was paved over for a highway, and the solar pans were destroyed by a hurricane in the 1920s. Syracuse still boils salt, and still pays the Indians, but the million and a half tons a year are half the peak of the century past.

One of the Onondaga competitors was salt from Kanawha, a valley in what would become West Virginia. Kanawha salt was red, too, tinged by iron and strongly flavored. The salt entrepreneurs of Kanawha tried, and eventually succeeded, in drilling through hard shale layers into an underground salt-water aquifer (drilling, coincidentally, through veins of coal). Tobias Ruffner was the man who made it work, doing what was thought to be a fool's work by drilling deeper and deeper. He worked his way below the weak red brine that others used and found a white brine ten times saltier. Kanawha eventually became the United States' first monopoly, an association of producers assigned production quotas based on the changing market value of salt.

Salt hurt the rebels in the War of Independence, when the Redcoats destroyed every salt plant they could find. In the same way Kanawha was a hotly contested site in the Civil War. The South had lost its transport lines and salt was dangerously low. The North took Kanawha and destroyed its plant and burners, and then the South captured and repaired it. But the North took Kanawha again and made West Virginia a Union State. The Confederates, like many other armies in history, couldn't continue without salt: salt to feed active men and horses, to tan leather for livery, to preserve food and make ammunition. Without Kanawha the South grew salt-hungry. Men who ran clandestine brine furnaces were exempt from the draft, but it was too little, too late.

I BUY TWO kinds of salt, a fine-grained iodized salt for the table (a delinquent behavior, I know) and a flaky, large-grained kosher salt for salads and cooking. Both are food grade, nearly 100 percent sodium chloride, and either could be native to the midwest, the coasts, a salt lake, or a salt dome. Most food grade comes from vacuum-pan production, a high-pressure method of boiling down the brine flushed out of deep salt veins—alas, untouched by human hands. A bit of food grade comes from solar evaporation, the coarser granules beaten down. My store-bought pickles might be steeped in solar salt, or rough-mined salt; my tap water softened by any of the three. Livestock eat rock salt, mostly, and the ice on highways is melted by it, too. Chemical processes take rock salt for the rough work, and vacuum-pan salt for the high-quality items. Any and all can come from anywhere, depending on supply, weather, demand, transportation costs, competition. Salt is still fairly cheap and still awfully heavy to cart around. But it's *present* in its most particular way, and I want to walk salt through, follow its path, and not just tilt the shaker. I want to find salt's origins.

My RENTAL car is gray with powder before I cross the dirt lot of a small salt factory in Utah. They make no effort to beautify the work; the building sits alone in an empty stretch of bare land, ugly with twisting pipes and silos and vents, surrounded by semi trucks and harvesters parked at random. (The harvester is a spidery machine perfected by a verbose and nostalgic man named Jim Palmer, who still works salt; it slices through a field of dried salt powder like a combine through wheat.) There are no parking spaces, just cars parked at random, as though abandoned in a storm. Scattered about are big piles of neglected salt, covered with soot and dirt. The powder floating about is salt dust, finer than cake flour and good for nothing.

This is a solar factory, crowded against the creeping edge of the growing Great Salt Lake, working acres of ponds. The flooding water has covered old ponds, invisible now, covered thousands of tons of unharvested salt not worth the price of digging yet. On the highway side the factory, to an untutored eye like mine, could produce anything: flexible couplings, zippers, dog food, saxophones. But out its other side is only salt, dammed in by dikes wide enough for a small pickup to drive on.

The ponds are a checkerboard, colored gray and green and a rusty red depending on the algae at work and the strength of the brine. (Red algae increase the heat, and heat increases the rate of evaporation; controlling that rate is one method for shaping the final crystal, square or jagged, in forms called stepped and hopper.) As the salty water gradually disappears a crust drops out, called bittern, formed of the other minerals that inhabit it—calcium carbonate, magnesium, and more. Different salts crystallize at different speeds and temperatures, and the process is such that sodium chloride is one of the last. The brine is pumped from pond to pond, growing saltier with each step up. Along the edges of the ponds sandpipers dip, searching for a breed of tiny, salt-loving shrimp. Brine flies, living only in salty waters, comb the sides in a cloud. It is quiet at the ponds; no harvesters at work today, no people. An abandoned campfire, ringed with stones, sits by the far pond's bank. The gates are locked and chained; the steep mountains ring the water like the fire's stones. From sky to hill to water and back is a palette of color, hazy, hot, redolent with spice and biochemistry.

As the brine is moved toward the factory—toward noise and people and machines—it gets saltier, suddenly crystallizing into a wet, sticky mush still thick with impurities. This salt, a basal form, is loaded onto a conveyor belt, passing along a scaffold crusted with orange and white stalactites, circles of sulfurous salt. The mush drips, and dries, crystals big as gravel, and tumbles onto a small mountain behind the factory. A mountain of salt compacted by its own weight as hard as glacial ice, marked with truck treads and so white with glare that I have to wear sunglasses to climb it. No one's around; this salt seems fair for the taking, but it bites back. A million cubes of salt, a million million hard as granite, and pressed so hard my fingernail breaks when I reach for a bit of it.

When needed, a slice of the mountain is cut, and run through a dryer. (My guide tells me there are people, "health food nuts," who want their salt without processing; they come and buy a bag or two of wet salt before it hits the dryer.) From the dryer the salt hits a cooler, which tumbles the suddenly sandy salt until it is cool enough to be handled. The whole factory vibrates without cessation; the ceilings and walls drip a stinking pink liquid chaff. All about my head are rattling pipes and humming canisters leaking furnace heat, metallic noise, rumbling mineral drums. This business of salt is a dirty one, the floor littered with puddles of mud and sulfur water, the air and every surface I touch hot, dusty, dry. The lines of pipes and pounding missile-shaped drums, the shaky stairs that twist to accommodate an agitated compressor, the hard-hatted men with seemingly little to do but listen to the rumble—I find it hard to follow. Not like the quiet ponds, one into another. This is human work, and not so clear.

The cooled salt is separated, a trough here, a chute there for various purposes; riding a belt up or across depending on its destination. On every floor I pass great green vats of salt so hot I drop the palmful I'm handed. A quiet man with a small grin opens a series of windows for me, doors like peepholes for the butler to spy out, and inside is a rain of salt, thick to medium to fine, pouring feverishly past. I have no idea where it's going.

On one floor—the whole building shaking here—a rain of salt falls into compactors, to be processed into salt blocks for cattle to lick, different blocks laced with iodine and sulfur and calcium for added nutrients, colored blue, magenta, gray, and sickly orange. Another machine smashes the salt into wafers, like bread for Communion, to be dropped in water conditioners. A bit makes it to food grade, ground round and round ceaselessly until it reaches the desired fineness, lovely and soft and white as Snow White's cheek. The floor is littered with great cubes of salt, treasure no one seems to notice, for me to pocket. (Secretaries, I discover later, have drawers full of them for visitors; I finger my stolen jewels in embarrassment.) Standing in the pounding noise I ask my guide about his health, and salt. "Love it!" he shouts over the noise. "No problem with *my* blood pressure!" And he turns to hustle down another stair.

Everywhere, everywhere is the salt dust. One empty room on the top floor of the factory is layered with the dust that drifts in the wind made by motion on floors below. It is fine as insect porridge, fine as spice, efflorescent white soot of gleaming scintillas. My clothes, hands, mouth, hair are thick with it; it lies deep in my lungs and dissolves with my breath's vapor into my blood, a drifting blizzard of minced salt soft as fleece and balmy warm. I sweat a salt-rich saturated sweat, my palms shiny with it, my ankles inches deep. It billows in little clouds when I walk. Neruda wrote: "Yo soy este desnudo mineral." I am that mineral nude. Like the eroding pillar on the bank of the Dead Sea, captive in the minor motions of time.

PART II
APPETITE

9

TASTE

THE ORIGINAL purpose of taste is survival: to know with a single sip or lick whether a food is edible or toxic, nourishing or deadly, out of the enormous range of sensations the world offers. In the natural world bitter means poison, and sweet, fuel. Even one-celled animals move slithering away from the bitter alkaloids; our own love of coffee is a learned response, requiring careful testing and a certain reward. It is the dilemma of the omnivore, as Paul Rozin says, to seek new sources of food at the risk of sudden death—or at least discomfort. The hunter-gatherer peruses every tree, every bush, bug, animal, seed, every root, flower, bone, shell, and fluid, and wonders if it might make dinner. Taste is his guardian angel, his guide.

And taste is more: it is the source of profound pleasure to most of us, coupled as it is with memory. Taste is not solely a physiochemical event of molecules and ions and membrane polarization, though it *is* that—but also a circumstance full of tones and variations, each one of which quietly etches the tone of a future response. Even in a room shaking with machinery and coated with salt dust it's the taste I'm after—the taste on the tongue, in the throat, the smell of it and the flavor licked off powdered lips. Taste between the teeth, on the palate, taste cuddling in the belly and licked off salty fingertips, the taste on another's lips, taken with greed, and then the aftertaste: lingering and dreamy. Even in a shaking room blurry with the finest salt, so fine it dances off when I reach. Especially there.

How did hungry ancestors give way to this? At what point did we find enough safety as a species—enough plenty—to learn to eat for pleasure? When did satiation blur into appetite?

In all mammals, taste buds are in the mouth, and they are similar across species. (Not so in fish—there the taste buds bow to the environment, and cluster on the lips, head, and body.) The taste bud is apparently simple but mysterious. It is stimulated in ways still unclear, responds in erratic and unpredictable fashion, changes for unknown reasons. It is a thing whose existence is a state of flux, whose *purpose* is to change, and change again.

Taste buds are specialized cells that cluster in tiny bumps on the tongue, concentrated on the front, the folds on the side, and the circular groove on the back. (Go on, get a mirror and look.) The tongue is a chemical sieve, a colander for stimulus, full of minute holes in otherwise unbroken skin. Each bump, called a papilla (Latin for nipple), has up to eight buds, and each bud has between 40 and 60 cells. Every taste bud has a central pore surrounded by protective epithelial cells that seem, to one biologist, "like shingles on a roof." Out of the pores shimmy tiny extensions of nerves called hairs; they are receptor cells, held just below the surface by branches of the cranial nerves.

Taste buds *look* different from each other, and they are in fact different. For a long time it was assumed that these differences meant there were specialized variations in taste buds, each variety serving a particular purpose. Current thought tends to see the variations as stages rather than types, because the mouth is in a constant state of decay and repair. No one knows how long it takes a taste bud to live and die, but by all appearances it is a fruit-fly life: fast and full of activity.

A cube of water—if one could be had—only one centimeter on each side has 3.3×10^{22} molecules: 33,000,000,-000,000,000,000,000 separate molecules. If this cube could be reduced to one-millionth of a millimeter on a side—10 angstroms—it would still have 33 molecules in it. Taste is a neuromolecular event, in which single molecules stimulate single nerve endings in particular ways. But which molecules, and which nerves? One carefully titrated drop of sugar solution floods the tongue with 100,000,000,000,000,000,000 molecules. How does the earnest researcher follow this giddy course? Consider what happens here: I take a bite of scallops meunière, firm but tender, warm almost to hotness, coated in a velvety sauce of oil and butter, thickened cream and the pulp of fresh garlic, laced with flakes of kosher salt. As I bite into it I feel the layers of scallop come apart like columns of basalt and my mouth is filled with a sweet soft salty taste of ocean. My nose fills with its scent, and a tapestry of detail awakens below my conscious conversation and awareness—of other dinners, other companions, storms, wood smoke, sandy feet, sea foam. I follow it with a splash of fine Bordeaux, astringent and sweet and sour all at once, clean and fresh. I haven't asked, but all my fine little taste buds have leaped to attention in a fraction of a second, and a world arrives. The spices exert themselves, early and late, dominate and surrender and marry, the flavors change, mold, and evolve, furrows rolling from the deep.

What exactly has happened? It is all a mystery, no more or less mystifying than the other senses, the Krebs cycle, the steadily multiplying zygote, the careful paring out of nutrients and wastes. For a long while it was assumed that taste was a reaction of enzymes, and that the molecules being tasted were actually broken into new compounds by molecules in the taste buds. But adsorption theory, which holds true for many responses in a variety of membranes, is currently the front-runner. The theory states that certain molecular substances in the taste pore change shape and size when molecules-of-taste (called "tasteable" molecules) are added. Not broken down but *conformationally* changed. These changes in shape and size then lead to changes in the electrostatic field, which excites the nerve. The adsorption theory accounts for the extreme flexibility and complexity of experience and explains—in a limited sense—mixed responses, like salty-sweet combinations.

The "phrenological" view of the mouth is still remarkably

common—that the tongue is divided into sections limited to sweet, sour, salty, and bitter. While fundamentally true, it is nevertheless inaccurate. Parts of the mouth and palate are more sensitive to one taste than another (although the "four tastes" themselves are widely dismissed now) but every area of the mouth responds to every taste. The degree of response varies, but taste, every single taste, is a whole-mouth experience. The palate, pharynx, and larynx are also taste-sensitive. Certain areas of the tongue secrete proteins that appear to be sweet-sensitive and bitter-sensitive, but there is no evidence that the nerve that tastes salt is any different from the nerve that tastes sweet. Interpretation occurs in the brain, and no one has any idea how.

Taste perception changes, too. A certain fraction of it is genetic—blacks show a consistently higher preference for salty tastes than whites, when all else is equal. Taking oral contraceptives over a long time increases the desire for salt. You can adapt to specific and singular tastes, grow used to them, and carry it into combinations: if you adapt to salt, and then eat something salty-sweet, the sweet tastes sweeter. If you adapt to a certain concentration of salt solution, a more dilute concentration is either sweet or bitter-sour. If you eat something sour close on the heels of something salty it can taste like water—not like nothing, but the almost-but-not-quite-nothing taste of water. Drinking water with even a small amount of salt for a long spell decreases a person's sensitivity—like becoming tolerant of a drug—so that you must eat more salt in order to be aware of it. At this point the research findings get really complex. One study compared a group of people who added 10 grams of salt to their food every day with a group who took the same amount of salt in tablet form, with none on their food. The group able to taste the salt came to have a higher preference, or desire, for salty tastes. Does the *perception* of taste create an appetite—through a growing familiarity, perhaps? Or does the sodium chloride react somehow with the taste bud, signaling a dietary change? Certainly there are chemicals that affect taste receptors and change perception; artichokes have a chemical called cynarin in them, which makes every taste following turn sweet.

And for all this I still want to understand that intricate connection between tongue and brain, that erotic relationship between appetite and thing desired. So a low-salt diet, after a few tense months, reduces the desire for salt. So a high-salt diet can increase it. The explanations being thrown around are less than helpful to the gourmand: "The surface pressure of charged phospholipid monolayers is a nonmonotonic function of the ionic strength of the solution subphase," writes one scientist, trying to make sense of the oddities. Perhaps it is, but still I wonder why. In all this talk no one has told me how I taste salt or sour or water. The nerve starts awake and the wave hurries along and breaks in the brain with a miniature and resounding splash, and I know. But *how* do I know? And knowing salt, *what* do I know?

It is tempting to presume that the enormous variation in molecule shape accounts for our differing perceptions. But a number of substances of extremely incongruent shape taste almost identical: sugar is sweet, and so is glycerin and chloroform. Many salts are as bitter as quinine and strychnine. (Again, those dangerous bitters. We are 10,000 times as sensitive to bitter tastes as to any other. Even coelenterates, those gorgeous primitives that include in their number jellyfish and coral, shy from bitter. We humans train ourselves to love them.) The chloride salts commonly taste salty, but potassium chloride and other potassium salts are bitter as well. You can taste substances—experience the taste on your tongue—given intravenously, and even certain solvents rubbed on the skin, as with the distinctive garlic flavor of dimethyl sulfoxide (DMSO). Rub a patch of skin on your ankle, and a few moments later you taste garlic—you even *smell* of garlic. (Rub a little DMSO on someone else's ankle, without wearing a glove, and you'll both taste it.) The combination tastes of salt-bitter and sweet-sour can be created purely by electrical stimulation of relevant parts of the brain, without any molecular presence at all. Perception differs in time, too: there is a normal circadian variation of the ability to taste a particular substance, or acuity. Researchers discovered some time ago that subjects who could recognize a specific minimal level of a substance in the afternoon couldn't recognize it the next morning. Acuity is always lowest in the morning, generally speaking. Perhaps this accounts in part for evening dinner hours.

The nose is intimately involved in taste, for several reasons. It is far more critical than the tongue, able to distinguish hundreds of specific odors. (And canine noses are many times more critical than human.) At the top and middle of each nasal passage are cilia, tiny hairs that float in the mucus lining. (Another aside: A scientist writing about food described mucus and saliva as "the body's portable version of the sea," because the two substances together dissolve molecules and transport them to their shore—the cell membrane.) These cilia contain receptor cells that can bind to volatile molecules—molecules which evaporate and rise rapidly. These olfactory cells are actual nerve cells, transmitting messages directly to the olfactory bulb of the brain, and from there to the hypothalamus (center for thirst and hunger and much more) and the hippocampus (a center for emotion). When food is eaten volatile molecules move into the nasopharynx, the passage from the mouth to the nose, and fit into particular cells by shape. Unlike taste buds, olfactory cells seem to respond specifically to molecular shape, the three-dimensional arrangement of atoms in balance. These shapes create odor patterns. So for each taste of a substance, there is likely to be a concomitant smell, and we are rarely able to separate the two. Though we say it "smells good and tastes better," in reality we smell and taste at the same time, and the whole is greater than the sum of its parts. Unfortunates who lose their sense of smell almost always lose, for good, their appetites. Food becomes a series of textures and temperatures, without attraction.

The trigeminal nerve is a tangled thing, stretching from the brain stem to the face, with branches in the gums and teeth, the nasal sinuses, the cheeks, lower lip, and jaw, behind the eye and in the cornea. It gives tactility to the surface of the eye, the mucus membranes, gums, and teeth, so that we feel touches there. It helps the mouth to chew, by stimulating muscles. The trigeminal nerve is truly the nerve of prickly things. Not only does it cause the occasional victim excruciating pain in Bell's palsy, but in the mouth it is responsible for our normal sensations of "bite"—pungency, sharpness, carbonation, and the piquant heat of chilis.

For a long time it was supposed that the trigeminal nerve endings in the nose could only respond to pungent and irritating molecules. Not so. The nose can be stimulated by purely volatile molecules, too, tiny clusters of atoms as smooth and unhurried and downy-soft as the odor molecules of a rose. People who have no olfactory perception, no ability to smell at all, can identify the rose by the *sensation* of the odor, as detected by the trigeminal nerve. What does the smell of a rose feel like? Does it curl up, slide, trip, or fly? Is it light, rich, strong, or dark? Whatever it is, a smell-less person can hold a rose to her nose and not smell it, but feel its smell.

Another possible influence on taste is saliva. Saliva is always salty, with a great variation in the amount of sodium and chloride ions between individuals. This creates a peculiar dilemma of taste research: Does the subject taste the sample or his own saliva? Saliva itself is tested for content. Stimulated saliva (created by three minutes of chewing rubber bands) has up to five times as many ions as resting saliva. The average adult—that elusive creature—produces about 1,200 milliliters of saliva every 24 hours. And young adults secrete significantly more sodium in their spit than older folks. But what accounts for its orgasmic position in the scheme of things, for the aphrodisiac flavor, that subtle, salty taste of mouths themselves?

THE DIZZY AND SUBTLE tastes of adulthood, composed of musky body fluids and complex mixtures of food, habituated bitters and deliberate sours and cloying sweets—these are as much nature as nurture. Even returning to the origin of taste, to the fetus swimming in fluid salt and sticky waves, we find the inexplicable. A baby not quite one day old can distinguish differing concentrations of sweetness, express preferences and dislikes for different tastes. The baby has had no sweet taste till that first mouthful—has, in fact, been drinking a protein-rich salty broth since conception—but it seems to recognize the worth of sugars. It is almost as though the baby remembers.

For nearly a century, researchers have been surprising infants with unpredictable fluids, gauging the nature and strength of their reactions, and mulling over the results. In the early 1900s such experiments were commonly done on severely defective babies in their first hours of life, in the hopes of discovering an instinctive taste center independent of a functional higher brain. (Infants with severe brain damage are still in demand for these reasons: scientists speak of "unique opportunities" and "rare chances.")

Give a newborn a bottle of sweet and it slips into a slow suck, steady, pulling the liquid in with the front of its tongue, protruding the tongue at the end of each suck as though searching for more. Its heart rate slows, too, and the face relaxes, yielding. This reaction is consistent and universal; it can be evoked with a single drop of sugar water.

In contrast to the reliable sweet response, infants push salt solutions out of the mouth, pause in their sucking, and make expressions of displeasure and rejection: tightly closed eyes, features screwed into a grimace. There is an increased frequency of choking with salt solutions, and the breathing pattern is disrupted. The infant grows restless, and moves its arms and legs. With bitter and salty solutions, some infants less than a week old will attempt to avoid the nipple. But even when actively drawing away the very young baby seems unable to stop sucking; the instinct to root for the nipple and suck overwhelms its discomfort.

In 1963 a hospital technician accidentally added salt instead of sugar to a batch of baby formula, which was then fed to fourteen newborns for several days before the mistake was discovered. The infants took their feedings with no more complaint than expected from newborns, and no one suspected a problem until the infants became critically ill. In fact, the babies developed a craving for the bottle, dire appetite and thirst, and demanded to be fed until they had convulsions and lost consciousness. Six of the babies died.

Oddly enough, a different phenomenon has happened to the babies of Moroccan nomads, who are used to a high level of salt in their natural water. The babies can grow deathly ill in hospital, malnourished and anorexic, as though dehydrated. They suffer a lack of salt when fed standard formulas.

The newborn's physiological need for sodium ranges from 6 to 8 milliequivalents every 24 hours. It so happens that normal breast feeding provides between 5 and 10 milliequivalents in that period. (Cow's milk, in contrast, provides three times as much.) Theories explaining why infants seek sweet foods and avoid salty foods rely on the dual qualities of breast milk, which offers quick energy in the form of high, easily digested calories, while providing the right amount of sodium to boot. Newborns have immature kidneys and a delicate fluid balance; high salt loads are dangerous for them. Babies are altogether ready to metabolize sugar, though; they release small amounts of insulin, a hormone necessary in the metabolism of sugar, with oral stimulus alone. Babies fed intravenously or through anesthetized mouths are unable to do this, and so don't benefit from this kind of anticipatory nourishment.

Experiments comparing infant response to sweet and salty tastes, however, have a peculiar bias: the salt is always given in a liquid form, and usually at room temperature. The same is often true of child and adult taste research; a dislike of a salty fluid is presumed to be a negative response to the salty *taste.* But with the exception of soups, salt is rarely taken in liquid form; few people find it palatable. Perhaps it is a memory of amniotic fluid, a dislike of one's roots and a desire to move on. Regardless of the meaning, it skews results. The taste researcher Gary Beauchamp compared the response of pre-school children to plain, moderately salty, and very salty water and soup. The children heartily disliked the salty water and greatly preferred the very salty soup, which had *twice* the concentration of salt typical to canned soups. Perhaps texture and temperature, and the multiplicity of flavors in a good soup, are as important to the appetite of the very young as they are to the gourmet.

Researchers become curiously nervous when they try to explain the baby's experience, and seem to find it much easier to discuss displeasure than pleasure. The adult context of pleasure often involves memory, and not uncommonly a memory of intimacy, of relation. Displeasure is easier to hold at arm's length; to admit that a baby feels some kind of sensual gratification requires admitting either that a baby makes memorable associations, or that nature provides for delight. Consequently the scientists trip over themselves with apology, none wanting to be the first to state the obvious: that babies show a preference for sweets because they *like* them. Writes one discomfited professor, almost in embarrassment, "I think that the infant slows down its sucking rate to savor the substance in its mouth. . . . I think it is a joy response, realizing that that is not good physiology."

This exclusive preference for sweet tastes tends to disappear around four to six months of age, when the baby starts to like salty foods—but again, in the form of food, not liquid. A two-year-old will reject salt water out of hand, but choose heavily salted food over unsalted food. The salt threshold of children—the level of concentration at which a person can taste salt—appears to be similar in both children and adults; the differences vary between individuals rather than strictly by age. Some children will eat plain salt to the point of choking on it, and ask for more. This growing tolerance, turning to desire, doesn't peak until adulthood, when a full third of all the sodium chloride we ingest is added from the shaker at the table.

In 1956 the scientist Hans Kaunitz published an article in the prestigious journal *Nature* called "Causes and Consequences of Salt Consumption." In today's positivistic biology, confident with its clutch of facts, Kaunitz may have cause to regret his essay. In it he strayed into the uncharted, the unknowable, beyond appetite. He proposed that the consumption of a greater quantity of salt than is physiologically necessary— what appears to be an addictive appetite for many—is the result of salt's ability to cause "emotional stimulation." Although 31 years' hindsight can make Kaunitz's cellular biology seem simple, his conclusions are surpassingly obvious to old salt eaters like me. He connects sodium chloride metabolism to the adrenal glands, and from there to the metabolism of sugars. He describes various conditions which are favorably improved by the administration of salt—like Addison's disease—and concludes that salt's efficacy "must be rooted partly in some pharmacological effect of the substance. . . . The stimulating effect of salt probably sets in motion adaptive mechanisms involving enlargement of the liver, kidneys, and adrenals. . . . Whether this is one of the roots of the reverence which was accorded salt by the ancients can scarcely be guessed at this time." Kaunitz goes so far as to wonder if salt improves learning ability, and, with a bow to low-salt diets in some diseases, concludes: "The physician, however, is not primarily interested in the mere metabolic processes but in the general welfare of his patients, and he should consider that the quickened pace of a more complicated society demands persons with a heightened responsiveness. Salt may be one of the ingredients producing this effect." What a sign of relief and comfort I heaved after finding this article! Scholarly, clear, rational, yet admitting of the most irrational of hungers. Is it stress that leads little fingers to the potato chips, the hot dogs and popcorn they crave? "Can that which is tasteless be eaten without salt?"

asked Job, the asker of good questions. Certainly not when times are tough. Then one gets one's salt however one can. (Certain African tribes get it from drinking cattle blood.) If Kaunitz can peruse Cro-Magnon cave art and imagine what the artists might have been if they'd had salt, then I can crunch a chip guilt-free, and look askance at my less stimulated peers who take more care.

At any rate, food preferences are highly contextual, and most particularly so after the physiological need is met—like any appetite, really—and with the possible exception of newborns, people can ignore their physiological needs to satisfy their appetites to a remarkable degree. Still, satisfaction actually influences appetite. In the phenomenon called alliesthesia, the pleasure of feeding an appetite—any kind of appetite—diminishes as the related physical need is satisfied. A food is better when we are hungry for it, literally, when the body's internal mechanism is needy. Like thirst, or sexual desire, or sleepiness, hunger is partly a product of poverty, and the glee we take in its satiating is a product of that, too. And an object once desired with utter abandon eventually becomes an object of dispassion—and then disinterest—as the body's needs are met. If eating begins with craving, it can eventually end in aversion.

This same temporary aversion can, with surprising ease, become an extraordinary psychological block. Many developmental models indicate that learning is most "skillful" in the early parts of an organism's life—most efficient, flexible, and amenable to guidance. At the same time, development appears to have critical periods for the learning of behavior. Introduction of a stimulus at a critical learning period under pleasant circumstances can create a lifelong desire for that stimulus: in other words, feed a child salty chips on a summer afternoon, follow them with a splash of lemonade, and that kid's hooked. But children as young as three and four can also express considerable disgust at the notion of eating certain things—things they have never seen an adult eat, or that have never been presented as food before, like insects. It is culture more than any physical appetite that directs our choice of foods.

Aversion—and preference—can be quite distinctive. If an animal is allowed to drink one stimulus—say, fresh water—unmolested, and then is shocked when he drinks the same stimulus in combination—salty water—he will learn only to avoid salt. The animal is able to distinguish the two components, sodium chloride and water, and to distinguish not only the taste of the components but, in his world, the significance of each.

A physical step beyond aversion is dysgeusia, an actual disturbance in the taste system. The most common form is hypogeusia, which is simply a diminishing of sensation (except for that annoying taste of bitterness, the one our very cells avoid). A quarter of all hypogeusic patients have high blood pressure—and remember the salt addict's need for an extra pinch or three? Hypogeusia is usually a product of aging, medication, or brain troubles. A hard fate, like smell-lessness, but preferable to its kin.

People with dysgeusia describe an awesome variety of sensations, almost all unpleasant. Some people hallucinate taste when nothing is there—one man tasted nothing but rotten garbage in his mouth, day and night. Others taste everything as bitter, or rancid. Still others switch tastes, like trains switching tracks, so that coffee tastes like shrimp and shrimp like cold spinach. The experience is crazy-making in the extreme, an unavoidable, invisible trauma that fills a man's days until he chews out the inside of his cheek, sucks his gums dry, worries the roof of his garbage-coated mouth with his tongue till he could scream. Dysgeusics have been wrongly diagnosed as schizophrenics in the past; a few, I'm sure, have gone legitimately mad. There is no cure except to root out the cause, often an unknown. A few have been saved by supplements of Vitamin A or with zinc, others with placebos.

Cancer causes dysgeusia, but mostly as an aversion. A form of this is part of the treatment: patients who grow sick after a chemotherapy session often develop strong and lasting aversions to whatever food they ate last before the treatment began. Most cancer patients have some difficulty with the flavors of meats, coffee, and certain spices, always tasting bitterness. Then there is the utter loss of taste of any kind, called, with a kind of poetry, "mouth blindness."

Salty tastes in the mouth are unusual, but they happen. (Better than garbage, I would imagine.) Inflamed gums and teeth can leave a constant, unrelieved saltiness. Lastly, there is simple indigestion, known more poetically by the old-fashioned term *waterbrash*. The backflow of acids from the stomach up can cause a most specific salt flavor that can't be erased by drugs or supplements or anything except the constant chewing of something else.

A WIDOWED king, with three maturing daughters, began to feel his own mortality bear down upon him. He wanted to determine how to divide his wealth. He gathered the girls together and asked each in turn how much she loved her father.

"I love you as much as I love my own life," answered one sweetly.

"I love you," said the next, with a look at her sisters, "as much as I love God."

The king was pleased at such devotion. He called his youngest, his dearest, to him.

"And how well do you love me, darling?" he asked.

"I love you, Father, as much as I love salt."

In all the stories, in Sweden, Rumania, Spain, Germany, Italy, and in England where the tale is called "Cap O' Rushes," the king throws his dearest child out in a rage. Like salt! A scandal! And the princess, in rags but forgiving and diligent and honest, makes do. Perhaps she eventually is taken as a servant girl, back into her father's kitchen, disguised. Perhaps she marries a prince who sees her true worth and, under a false name, invites her father to a banquet. Either way, she conspires to cook for him. At a happy feast she serves him a secret dish, like all the others but minus its precious salt. ("That will be rare nasty," says the cook in one version, when she proposes the experiment.)

The king can be polite for only so long before he leaps from his seat in disgust, watching the other diners in dismay. The food is awful! he exclaims. How can it be eaten! And the princess reveals herself, kneels at her father's feet.

"I tell you again, my father," she says, "I love you as that most precious thing, as much as salt." And he understands.

THE CYCLIC nature of ideas is revealed in the history of taste theory. The theories presented have repeatedly evolved from the simple to the complex to the simple again, and the cycle shows no sign of halting. Such distress of theory is understandable: Taste *is* simple, and infuriatingly mysterious at once. I know why I stopped this morning to buy a bag of potato chips—what explanation is necessary? I like them. But there's so much more. Why this morning and not yesterday? Why potato chips and not chocolate? Why, for that matter, food? Why didn't I take the urge for motion and pleasure and turn it into a tennis game, a side trip to the museum, or the purchase of a new coat? All these things serve a need. But transmutation is absurd when it comes to appetite, doomed to failure and an obsessive memory of the not-done. When I want potato chips that's all I want, and all you want when it's your turn; the salt, yes, and the bite, the carbohydrate fuel and the smooth oil, muted blend of flavors with aggressive textures and shape. I find it a little silly to try to differentiate the many flavors of a chip, any more than those of a scallop in sauce—it seems a tautological quarrel, because a chip and a scallop simply taste like themselves. But the theories fascinate, attempts as they are to imagine boundaries where none seem to exist, to define the indescribable, especially to set limits on the illimitable, as though a wholly new flavor discovered tomorrow (like the happy mating of peanut butter and chocolate, perhaps) would be disallowed under the rules.

One of the first theories of flavor was proposed by Aristotle around 530 B.C. Nothing escaped his scrutiny. He suggested seven flavors: sweet, bitter, sour, salty, astringent, pungent, and harsh. (A mere 600 years later, Pliny the Elder said, "Even the pleasures of the mind can be expressed by no better term than the word salt.")

Several centuries later the Russian Mikhail Lomonosov took Aristotle's mantle and advanced another seven-flavor list, with examples of substances that expressed the flavors precisely: acid (vinegar), caustic (grain alcohol), sweet (honey), bitter (tar), salty (salt), pungent (wild radish), and sour (unripe fruit).

It was not till 1916 that the German scientist Henning developed a theory of flavors that attempted to account for the multitude of combinations that most people experience. Though Henning's theory is often referred to as a theory of "primaries," what he proposed was a tetrahedron—a four-faced pyramid—with "innumerable transition points." At each of the four points of the tetrahedron is a "primary" flavor: salt, sweet, bitter, and sour. Along each of the edges are continuums of two-flavor combinations: salty-sweet, sour-bitter, and so on. Each of the four faces is a three-flavor combination: salt-sweet-bitter, sour-bitter-salt, and so on. The result is a simple, transparent, hollow structure: hollow because Henning felt no four-flavor combination was possible.

Henning's geometry works in that it allows for almost endless variety. He compared the flavor hues to hues of color— that sweet-salty is to sweet and salt as orange is to red and yellow. Orange is not red, not yellow, not a wavering border between the two, but a "uniform sensation of fusion" between red and yellow. Furthermore, to Henning combinations were *nonreducible*, the way the taste of the scallop cannot be broken down into simultaneously occurring quantities of shellfish, salt, heat, moisture, and firmness. As orange can't be turned into red and yellow without the loss of its "orangeness," neither can a food both sweet and salty be split without its disintegration.

But Henning's structure also fails, for two reasons. Unfortunately, time has shown that a number of substances fall out of its boundaries, at least as defined by research subjects. For me the limits of its elegance are its downfall: things taste different from day to day, moment to moment, dependent on what precedes and follows, and more. Would Henning have me place my first, unique, nonreducible bite of scallop *here*, just so, on one face, and my second, equally unique and non-reducible bite over here? As an explanation of taste experience, Henning's tetrahedron is a guide—but not enough. Besides, where in all this do I place the lingering aftertastes, and—much later—the memory?

Gradually researchers (in frustration, perhaps) have grown less interested in the perception of taste and more concerned with the cellular level—the chemical event itself. Lloyd Beidler, in 1954, proposed a complex theory of adsorption, or binding, still in force today. It is a microscopic, and therefore microcosmic, view, in its essential form a mathematical exploration of thermodynamics, and a far cry from talk of "flavors" and "hues" and "experience."

Beidler's theory has several parts. He said first that stimulation is independent of time—that is, that it continues at the same intensity as long as the stimulus is applied, and doesn't drop from an initial peak. The neuromolecular response is completely reversible, and both parts of the molecule—the positively charged cation and the negatively charged anion—are involved equally. Beidler felt that, even though the stimulation continues through time, a plateau of response is reached at some point, when an *increase* in stimulus won't increase the response. Beidler was the first to say that tongue receptors respond to many different substances over a wide range of concentrations. His theory was radically different for what it was not, as much as for what it contained. It diverged from enzyme theories—substances "breaking down"—to adsorption theory—substances "building up." In either case the change in polarity stimulates the nerve, which leads to perception. (Whatever that may be: here the visionary Beidler stops.)

In vogue now—the latest in the cycling of simple and complex—is multidimensional experience. Current thinking holds that (as most gourmets would agree) the distribution of taste receptors, regardless of the lack of certainty, is irrelevant: taste is a whole mouth-palate sensation. Combinations of three or more tend to have a canceling effect, and one component tends to take precedence. But two-flavor combinations can be remarkably satisfying—certain combinations. The most pleasurable are sweet and sour, lemonade or a toss of pineapple and vinegar in sauce; sweet and bitter, the supreme example being chocolate; and the most common is salty and sweet, the purity of fresh baked bread. These two-flavor combinations fuse; the others—salty and sour, salty and bitter, bitter and sour—don't, remaining distinctive and unpleasant, like battling neighbors. According to a multidimensionalist, the most successful foods are those which offer flavors following each other in rapid succession, rather than purely in fusion, giving the mouth a rollicking series of tastes in an unrepeated pattern. This succeeds exactly where Henning fails, with a bow to the transient perfection of flavor, and its contextual nature.

A lot of us are multidimensionalists without knowing the word; it is flavor succession and the tapestry of change that we seek. M. F. K. Fisher writes of a man who loved to eat heavy Russian rye bread. He would spread the bread with thick slabs of sweet butter and "then, to my private horror, coat the whole with an impossible load of table salt; he likes the odor, texture, taste; it makes him feel good to eat this honest, enriching fare."

JAPANESE cuisine is full of multiplicity and changing, shifting boundaries and categories bordering on the individual. As strongly as Americans recognize sour, sweet, bitter, and salty, the Japanese recognize *umani*. It has been described in pathetic inadequacy as "Chinese food taste, meaty, or bouillon like," but is most commonly referred to as the taste of MSG. But there is more to it than this.

Japanese cuisine makes frequent use of broth for seasoning, and three common stocks are made from kombu, a kind of seaweed; katsuobushi, derived from the bonito fish; and shiitake, made from mushrooms. It was from kombu that MSG was originally isolated in 1909 (it's now made commercially from fermented sugar cane), but katsuobushi and shiitake also share the *umani* flavor. Japanese research has not only isolated the free molecules in bonito and shiitake that give the flavor, but determined the amounts of these chemicals in other foods. (Parmesan cheese, dried sardines, and fresh tomato juice all share it in quantity.) The commonality is amino acid, and as salty flavor usually signals the presence of sodium chloride, the *umani* flavor signals the occurrence of amino acid. It is no accident that the Japanese salt substitute ornithyltaurine is an amino acid, too.

Monosodium glutamate (MSG) really is a singular flavor. It is a salt of glutamic acid. In time the Japanese learned to make it from wheat, and by the 1950s from a bacterium that leaves glutamic acid as a waste product. No one knows how MSG works, but it undoubtedly enhances other flavors as well as providing a saltlike *umani* flavor of its own. In large quantities on an empty stomach, MSG causes allergic-type reactions; some people are quite sensitive to it, giving rise to the occasional case of "Chinese restaurant syndrome," with symptoms of headache, sweating, an alarming chest pain, and burning skin. It is a salty burn, the rough rubbing of skin and the tightened muscles; a wet, hot overdose of spice best taken in small measure.

IN ALL THIS talk of flavors, how many, what kind, and examples—this ultimately nonsensical attempt to delineate one of the more subtle and provocative of human experiences—certain questions can be forgotten. One is the point of primacy, a return of Henning's infinite combinations, like points on a line. Rather than ask how many single flavors there are, one can ask whether any flavor at all is primary—that is, solid, singular, unique, and not reducible to parts. Is a taste—any taste—separate and complete in itself, or a flashing series of changes? What, for instance, happens when you take a bite of perfectly softened, cold ice cream? There is, in that brief moment, an enormous number of sensual offerings: texture, temperature, of course, and a curiously complex flavor that is *as much* a product of the former as of the flavor components. But how do we describe ice cream? Sweet? Cool? Vanilla or chocolate? The language is strikingly impoverished. Quinine, morphine, strychnine, caffeine, and urea all taste different, test subjects agree—all unique, unquantifiable. No one is able to suggest a word for the taste of any of them, but *bitter*.

In comparing the experience of taste with seeing and hearing, Robert Erickson, a psychologist, gives a sense of the difficulty vocabulary presents. It is Erickson's feeling that the whole concept of taste is somewhat misdirected. "Aristotle seems to have set the mold," he writes; "by posing the question in terms of *how many* tastes there are, he seems to have largely prevented the question of *whether there are* primary tastes at all." Erickson has worked with hearing and vision in ways analogous to taste research, partly in an effort to determine whether unpracticed test subjects have a "natural" ability to discriminate difference. With hearing, he adds, "there is absolutely no difficulty" in people knowing whether they hear one or more than one tone. Likewise, with color, subjects find the question almost silly: "I have gone so far as to jeopardize my reputation as a responsible and meaningful scientist by asking subjects whether each of a variety of hues [of color] was one or more than one; the answers were always unequivocally 'one,' even for colors intermediate to red, yellow, green, or blue. No practice was required. The subjects were puzzled that we should ask such a simple question because all hues are so clearly seen as 'one.' "

When Erickson began to test taste, however, response was much more unpredictable and varied. It was not clear if subjects tasted different primaries or "side tastes"—like aftertastes. Erickson's experiments show that no one taste was completely unified. Even tastes with side tastes, or that cause complex reactions in subjects, are not clearly more than one taste. Each taste seems to present itself with a kind of dignified uniqueness; each taste is simply *itself*. The long-accepted taste "standards"—like sodium chloride for salty and quinine for bitter—may be more clearly delineated as primary tastes, but even these pure samples shift and shimmer with impressions of other flavors, other tastes, as a drape of peach-colored fabric will shimmer in light, first beige, then pink, then oddly orange, but never to be described as anything but peach.

(Sodium chloride as a standard: it makes sense. Such a simple molecule and so full of vitality. Salt has that requisite dignity, and it has flight. Salt stays, dry and soft; it moves from a quartzlike durity to the breezy dance of dust. I can hear salt laughing at our efforts to pin it down.)

For now, one of the most acceptable experimental methods for charting the differences between tastes—if not for finding agreement—is the use of scaling maps. These are pictures of a taste geography, as it were, two-dimensional representations of a variety of flavors in relation to each other. Research subjects simply map a long series of tastes according to the degree of similarity and differences, placing each close to similar tastes and far from different ones. So a relatively "pure" sour will be far from a relatively "pure" bitter, while the bitter may be closer to certain kinds of sweet. With scaling maps different subjects find more room for agreement than disagreement, partly from giving up a dependence on adjectives, and at the same time the entire knotty question of combinations versus primaries is avoided. One interesting point found with scaling maps is that the greatest predictor for a salt's placement on a scaling map is not its positively charged ion—like sodium—but its negatively charged ion, like chloride. Chloride brings sodium and potassium and magnesium close; but sodium fluoride, sodium acetate, and sodium bicarbonate may occupy greatly divergent positions. (And each of the latter may be far from sodium chloride and the other chloride salts.) Why? What is it about electrical charge? What is it about the shape and vibration of the outer shell of electrons that makes for flavor? Is it that the anions like chloride define the character of the positive cations like sodium, placing them in categories the way clothes make a man? Or does sodium have a strength of its own, a multiplicity in its very structure that allows it to branch far and wide, pass borders, ignore the limits of a quieter structure like chloride?

Henning's tetrahedron really shows its limits when compared to scaling maps. This is the result of talking about differences rather than qualities—it allows a taste to oscillate in the mouth's wind, to occupy an area rather than a point. One researcher has found 13 sodium salts that fall outside the faces of the tetrahedron completely, a tetrahedron that combines all four flavors into apparently infinite combinations. One such salt is monosodium glutamate (MSG). Some experimental subjects say it is not salt, not sweet, not salty-sweet, not salty-sour-sweet, not comparable, not definable, except as itself.

ALL THIS TALK about definition seems rather pointless until the doctor says to cut out salt. (No one's said it to me, yet, in spite of my father's soaring blood pressure and my own salty predilection—but if and when they do, I think I'll go digging into the calcium research, look into that non-salt-sensitive sixty percent, check my options.) The most detailed descriptions of the adult experience of salt come from the demand for a perfect substitute, a surrogate for salt's ringing flavor. Potassium chloride (KCl), long the most commonly used salt replacement, suffers from bitter side tastes, a lack of strong "salty" flavor, and the potential for causing potassium overdoses in people whose cardiac health is already compromised. MSG has a number of drawbacks for health in certain individuals, in spite of its *umani* quality. And nothing but salt tastes salty in just the right way.

One experiment just attempted to find the particular *qualities* that distinguished sodium chloride from potassium chloride (KCl) and MSG. Each was described in two categories: "mouth feel" and "off notes." MSG had a warm mouth feel, and KCl an astringent one; common salt, on the other hand, was described variously as cool, smooth, and oily. Off notes, negative or unpleasant side tastes, for MSG included a metallic taste, bitterness, and chemical and soapy flavors. KCl was also metallic, bitter, and had a dishwatery or plastic taste to some. And even sodium chloride tasted metallic, bitter, and like dishwater. (In an attempt to find a word, *some* word that works, researchers continue to list adjectives: others used in the past for salt are sharp, tingly, sickly, bright, rough, tangy, soft, flat, clean, rounded, full, and thick.)

The research subjects made two unexpected points about salt. The good qualities of NaCl weren't limited to its own taste; the subjects also felt that sodium chloride in foods increased the sensation of "fullness and thickness" in the food, increased the sweet qualities of a food, and lowered the perception of chemical off notes. Sodium chloride affected the overall "flavor balance," and regardless of the degree of saltiness per se, improved the pleasurable quality of the food.

In the final analysis, in fact, the least essential quality of the seasoning used was its saltiness. The most important was its life on the tongue, its evocative motion, its nude wholeness of molecular structure and anionic charge and electrical velocity: the most important thing to a seasoning, and the quality in which sodium chloride was the undisputed champion, was mouth feel.

In 1925 a physician named Henry Green wrote an article called "Perverted Appetites," in which he discussed all manner of unusual cravings in human and animal alike. He described wool eating by sheep, feather eating by chickens, hair licking, coprophagy, the insatiable desire to lick and chew, the eating of the newly born, and salt hunger. In passing Green mentions pica, long defined as the craving to eat the inedible. It is, said Green, recorded "amongst hysterical females during pregnancy, and amongst idiots and the insane. . . . It seems an idiotic sort of thing to do and the physiologist is tempted to leave the psychologist to study the problem."

These somewhat more enlightened days give a craftier definition of pica, now recognized to be an organic kind of idiocy, a physical compulsion of metabolism that creates an uncontrollable urge to eat a particular thing—not always edible—and to eat it in the most outrageous and inexplicable quantities. Most pica victims eat starch, or clay, or chalk, and it is supposed there is often a missing nutrient at the root of the urge. Some people eat ice, others lettuce and raw potatoes, to the exclusion of other food, and to an extraordinary excess.

And salt. What of salt? Besides the hypertensive love of it, some cancers seem to make salt taste more palatable, or increase the desire for salt. Some plain folk like me, normotensive and ordinary, seek salt as a pig does a truffle. I think Brillat would understand.

There is a single case on record of salt pica. The young woman involved seemed healthy. She didn't have Addison's disease or cystic fibrosis or a tumor or kidney failure or any of the score of illnesses that could create salt hunger. She was slightly deficient in iron, slightly anemic. She claimed to be thirsty; her blood pressure was a little high. This woman ate a half-pound and more of salt a week, by the handful, and carried packets of it in her purse for snacks. She ate salt till her lips cracked with its aridity, and liked it. She didn't particularly want to stop, either, and one can certainly imagine worse habits. The doctors treated her anemia with iron supplements, and her hunger for salt disappeared. Now, I suppose, she's just like the rest of us, shaker at the ready, only half-conscious of what the appetite seeks.

10
HUNGER

THE LIONS, leopards, and hyenas were growing arrogant and unafraid. More and more often they crept close to the borders of Kumati's village. Kumati was a clever young man who had yet to prove himself as a warrior, and, listening to the soft pad of paws outside his hut one night, he thought of a way.

One night he crept away into the forest with only a bow and arrow and a bag of salt, and found a hollow beside a tree, and slept. Kumati woke with the sun. The face of the leader of the lions was close to his.

"Why do you lie there and wait to be eaten?" asked the great lion.

"I have come to be your servant," replied Kumati. "I will make you the most wonderful meals you've ever tasted. *I* have magic." And he held up the bag of salt.

The lion considered this crazy young human, so small he was hardly worth eating. "I'll give you a chance."

The next day Kumati left the beasts to drowse in the cool shade and hunted. In the evening he returned with meat, which he cooked over a slow fire.

"My magic won't work if anyone looks," he warned, and the animals turned their heads. Kumati sprinkled salt into the pot. The animals were greatly pleased with Kumati's skill, proclaiming his food the best they'd eaten.

One night Kumati made all his preparations the same as before, but left out the salt.

"This isn't like your other cooking," complained the lion. "It has no taste."

"I'm sorry, O Lion," answered Kumati. "The leopards peeked, and broke the magic."

In spite of the leopards' protestations of innocence, the lions and hyenas set upon them in frustration, and killed them.

The next night, Kumati prepared a delectable gazelle, but without the salt. Again the lion complained.

"Oh, I'm so sorry, Lion," said Kumati with downcast eyes. "But the hyenas peeked and the magic failed."

That was all the lions needed to hear. They set upon the frightened hyenas and killed them all, every one.

The third night, Kumati hunted a fat antelope and cooked it as before, but without salt. The lion was, by this time, hungry in his belly and anxious for the fine pungent flavor with which he'd been so quickly spoiled.

"What has happened?" he demanded after a single bite. "Shall I eat you, human, since your magic doesn't work?"

"O King," whispered Kumati with sorrow. "It isn't my magic. This time, *some* of the lions peeked."

And of course the lions set upon each other, forgetting their relation, and Kumati crept away to his village, victorious.

Aн, DEAD BY salt. It melts in the throat, cool and seductive, addictive siren. But enough doom and gloom. Let's eat.

Chips and dips, pretzels, peanuts, pizza, popcorn, and pickles. Let's eat. Salami and salt roasts and jerky—soda crackers and chutneys and sauerkraut. Let's eat anchovies, olives, bacon, and more. Eat—then drink, that's how the saying goes: "Salt beef draws down drink apace." The one makes the second so sweet. Crunch and bite, sharp tastes, soft corners, a friendly crunch, and then a bright sour splash in the mouth.

Aristotle liked it. The Greeks praised it. The Egyptians preserved with it, a long time ago, dead people as well as meat. The Spartans were fond of "black broth"—pork juices, vinegar, and salt. (Might as well eat.) And a *very* long time ago, say some, animals drinking melted glacial waters grew hungry for salt, and migrated to the sea to find it. The early humans learned—one wonders exactly how—that urine was salty and good to mammals. They wagged their flattened, big-browed heads, and thought, and thought, until they thought to bait traps with human urine.

The Romans, though, were fond of liquamen. It was more than salt, or rather an evolution of it. Liquamen was manufactured along seacoasts, in Pompeii and Antibes, a real industry—labor-intensive. To make a tub of liquamen (a delicacy in its finest form), one took a number of fish and mixed them with a large quantity of salt. The fish could be small or large, white fish or herring, or even shellfish. The mixture was laid in the summer sun in open earthenware vats for a spell: two months, three months, eighteen months sometimes. As solid turned to liquid and filled the basin it was drained out and poured in again at the top, the remaining fish were stirred, the aroma a vapor of sea and salt. The result, writes one food historian, was a clear golden fluid with a "distinctively salty, slightly fishy, slightly cheesy flavor," not unlike the nutritious fermented fish sauce used in Southeast Asian cuisine. The salt inhibits bacterial activity. It holds the fish, as it were, in a kind of suspended animation of decomposition—fluid, clear, unfinished.

Liquamen, because of the salt, could never achieve the complete cheesiness of a food made by the Inuit. Their saltless paste of decomposed fish heads, fins, tails, and guts transforms into a bland paste with almost no flavor of fish at all—with almost no flavor, period. The liquamen of the Romans decomposed to a peculiar level of fermentation, held in check by the power of salt to halt corruption, but taken there by the equally powerful effort of fish and bacteria to work decay. The tasteless paste of the Inuit decomposes even further, to a pure pickling of bacterial mutation, a curd of digested and excreted protein bound in a web of the microscopic.

All the forms are part of salt's centricity; where it has been known it has been essential. Once discovered, it remains. The Puranas, or "Ancient Stories," of India provide one description of the universe, a set of seven concentric rings or oceans surrounding Mt. Meru, the human plain. The outermost ring is water. The next closest is made of curds, then comes milk, then ghee, or clarified butter. The third ring is wine, and next to it a dark sugar. Closest to the human world is an ocean of salt.

ITS POWER IS political; salt's necessity is its power. Even the place names reflect this: Salamis in Cyprus, Salinae in Italy, the Saltcoats and Prestonpans of Scotland and the Via Salaria in Rome, Salins of Burgundy and all the different *wich*es of England, each a former site of salt production. Even salary is salty, a Roman derivation from a time when a soldier received part of his pay in salt, and hoped to be worth it.

The finest, lightest salt of the European Middle Ages was made by burning dried peat that had been soaked in seawater. Peat salt was actually white, a highly prized quality, the product of an involved process of dissolving the salt ash left from the burnt peat in fresh water and evaporating it. The result was a delicate and expensive commodity reserved for guests, in an economy that left many hungry and simple nourishment a prize.

The producers of peat salt competed with the makers of brine salt, who boiled down the water of salt springs. Brine salt often had impurities, and cooks exchanged suggestions for cleaning its flavor of the bitters and metallics that clung: egg whites, cow's blood, sheep's blood, beer, each was said to pull out the flavor. (This was, undoubtedly, a technical improvement; the early Britons merely poured brine over hot coals and scraped off the crust.)

The third competitor in the field was Bay salt (named for its source, Bourgneuf Bay, on the French coast, where it was first produced by evaporation), costing half as much as the other light salts then available. Bay salt was cheap not only because of the quantity on the market, but because no effort was made to purify it. The seawater was collected with dirt, seaweed, and flotsam mixed in, and the result was a rough and discolored grit, ranging in shade from black to gray to green. Bay salt was bad for preservation, so coarse it couldn't cover the surface of the food. It had in fact to be ground before it could be used, and the more leisurely households employed a servant for the purpose, called a powderer. The prudent household had more than one kind of salt in its larder: Bay salt, for cooking when no guest was watching, and brine or peat salt for the table.

For the most part salt in medieval Europe was made by Catholics, primarily because of their location along the hot coastlines. It was largely sold to Protestants, and in the worst of religious climes the trade prevailed, too important to be dismissed for merely theological antagonism. With the spread of the 40-day period of Lent the trade in salted fish grew even larger, and so did the coastal fishing towns.

The Eastern European salt trade was dominated by Jews, especially in Poland. (Jews were involved in Spanish and German salt production, too, and in Muslim countries.) Their control, in fact, became alarming to the nobility, who spent much time and political favor toppling the Jews from this throne. Anti-Semitism in Poland eventually grew so strong that no Jew could lease a salt mine. In 1824 a law was passed prohibiting any Jews from gaining new salt markets until the number of Christian traders equaled the number of Jews. The equality was never achieved, and the time came when the Jews were partitioned out, and the salt left to other hands.

The chemist Rudi Bloch has made a study of the changing level of the sea on the rise and fall of cultures. He claims, looking at Europe, that a free and democratic society requires plentiful salt. A commodity controlled by the government— like salt in Poland, tied to the failing power of the nobles—is a monopoly. A commodity that answers a metabolic need—be it salt, fresh water, grain staples, even firewood—is a monopoly on health. Its limitation becomes an enslavement. If the Catholics had been willing to lose their salt profits, the landlocked Protestants might have lost much control in the long antagonism of Christianity.

Salt withheld is slavery; in Africa salt bought slaves. The value of salt—and slaves—rose and fell in the tribal economies with supply and demand. A block of salt would buy a healthy young girl one year; the next year her price would be up to five. This is salt hunger turned to money by way of flesh. The bound girl kneels in front of the buyer and submits to examination, her dusty face pressed close to the dirty rough edges of a nugget of salt.

France had the infamous *gabelle*, a salt tax that literally priced the spice out of reach of the poor until they grew sick with the deprivation. In 1547 *la gabelle* became a cash tax, forcing an economy of barter to become, abruptly, an economy of cash. In 1548 a peasant rebellion temporarily halted *la gabelle*, but it returned, in the sixteenth century. Beginning in the early 1700s through the long reign of Louis XV *la gabelle* grew and grew to absurd heights, fueling a revolution.

Salt as money: it's been traded ounce for ounce with gold in the past, and made into money in the Himalayas and Upper Senegal. An Abyssinian saying defines a millionaire as "he that eateth salt." There people wore salt sticks around their necks, for food and money both, a precarious form of banking but right somehow. All the economics of salt depend on hunger, a cellular hunger, an organic appetite.

Among other subjects worthy of argument, Queen Elizabeth and Mary, Queen of Scots argued about salt. Elizabeth monopolized its production, limiting each *wich* to six days of boiling in the spring, and six days every fall. The noble purpose was to keep the price high, because salt was taxed.

Mary found an Italian who claimed a better method for salt making, and she issued him a patent, hoping to break into

England's market. Elizabeth then found a German with a similar claim, and tightened the screws on the market. Salt prices soared; many people couldn't buy enough for themselves, let alone their livestock. Smuggling grew on the English-Scottish border, and from Ireland to England, and men were imprisoned and hung for salt smuggling as late as 1785. Only a popular rebellion stopped the price and brought the tax to a tolerable level. (The United States wasn't immune; the first duty on salt was imposed in 1797, as a war tax. It was repealed and reinstated many times.)

In 1930 Mahatma Gandhi made salt an icon of freedom from the colonial yoke of Britain. The British imposed a salt tax. The law extended to the mere possession of raw salt by an Indian. People who lived on the seacoast were forbidden to pick up a single cube of it from the shoreline, or dry a bowl of ocean water in the sun outside the door. The Indians labored in the sun and had little variety in their diets; salt was as essential to their health as water. Gandhi, wrote one historian, "came to look upon the salt tax as a tax on Indian sweat and blood."

The march to the shore to make salt, beginning in March of 1930, ultimately became the largest protest movement in all of Mahatma Gandhi's enormous effort to bring independence to India. It began, after a walk of nearly 200 miles, when Gandhi himself entered the sea, bathed, and then began picking up raw salt from the shore. He was immediately hailed in joyous cries of "Law-breaker!" His example was followed by thousands of Indians for months, panning seawater, finding raw crystals, making and selling contraband salt across the country.

Eventually the police began to use violence, and make arrests. A demonstration of civil disobedience at the Dharasana Salt Works north of Bombay—a silent gathering at a time when assemblies of over five were prohibited—resulted in 320 Indians seriously injured, and 2 dead, from beatings by police.

Gandhi's protest against "the most iniquitous" tax created

a great popular revolution, a wave that swelled like the tide. The salt itself, wrote Nehru later, was nothing special. "We knew precious little about it, and so we read it up where we could, and issued leaflets giving directions, and collected pots and pans and ultimately succeeded in producing some unwholesome stuff, which we waved about in triumph, and often auctioned for fancy prices."

MEDIEVAL people ate between 10 and 20 grams of salt a day, far more than the average modern American. This was partly in the form of salt beef and salt pork, the only form of meat available to the working class in quantity. But it was also a free-sailing cuisine at work, with salt poured into foods for preservation, sucked out in cooking, then poured on again in preparation and at table.

Medieval tables mixed the flavors of sweet and salty almost whimsically: desserts were salted, salted meats and soups flavored with fruits and sugars. Eventually salt found its way onto the table, but spent time there almost as a decor rather than a seasoning; the food was already salty enough.

On the table salt was kept in salt cellars, elaborate contraptions of precious metals, often sculpted into shells, dragons, or ships as high as two feet. They became most important utensils, embodying not only expense and artistry, but the potency of salt's supposed magic, its ability to spread evil if mishandled, bring luck if treated with care. Cellars were symbols of wealth, status, and superstition. The salt itself was carefully piled in the center—perhaps in the hold of a perfect ship with eight miniature crewmen—and rounded into a small white mountain.

The grandest salt cellar was placed beside the host, or the guest of honor if he was of sufficient importance. Gradually a tradition emerged, a seating plan. Guests seated on one side of the cellar were honored. Those seated "below the salt," as it came to be called, were the common people, or intimates in the host's disfavor.

Good manners dictated that the expensive pure salt on the table be kept pure. Servants were exhorted not to scoop the meaty leavings off the table back into the cellar, lest they contaminate its color. A guest was expected to keep his fingers out of the salt, if not out of anything else, by scooping a portion out of the sculpted mound with the end of his knife. (All but the most important guests provided their own knives.) Plates were trenchers of bread at first, hollowed out; extra trenchers to hold a person's salt serving were available. Later, when wooden plates came into fashion, they often had small depressions for a scoop of salt.

When I was a child, all we had was a simple matched pair of glass shakers. Both were filled with fine, powdery commercial salt, one on each end of the table so that it was always in reach. We salted everything, before tasting—no matter that my mother salted the canned or frozen meals while they heated. "Pass the salt," we said to each other. "Pass the salt."

Now my father's table has five shakers on it in a permanent display. One has salt, and one pepper. Two have spice mixtures made for him by a concerned aunt. The last, the one he uses, has potassium chloride, and he sits heavily down to his meals, his blood pressure a red mask on his face, and mournfully sorts his condiments before he begins.

The exhortation not to touch the salt persists, and especially not to spill it. The Romans are said to have led victims to execution with salt balanced on their heads; when it spilled, so did their blood. Spilled salt means a broken friendship, a broken bone, a broken contract; it portends a shipwreck, danger, heartbreak, loss, and death. If salt spills toward you the bad luck can be countered by pouring wine in your lap—a distraction to the spirits, perhaps. Demons (*sinister* by nature) prefer the left side, so throw a pinch over the left shoulder to satisfy them. Judas spilled the salt at the Last Supper—or at least that's how da Vinci painted it in his time. "Never borrow salt," the saying goes. "If you borrow salt, don't return it." It's considered bad luck to run out of salt, spill it, waste it, loan it; to salt another's food, or to pass salt hand to hand. "Salt or brains, do not offer," goes one superstition. One should, instead, place the salt beside the person and withdraw one's hand, as though the salt had the power to pass a current of power through it like wire. The Russians had a saying of caution: "Don't pass salt—but if you have to, smile."

"SALT COOKS bear blame, but fresh cooks bear shame." An obscure warning, at first. But fresh food poorly preserved brought illness and sometimes death. Too much salt could get a cook in trouble, but in certain cases too little was worse.

Salt preserves in two ways: it withdraws water, and slows bacterial growth. In the diet system of macrobiotics, salt is *yang*, cold and strong, and water is *yin*; they balance in a dance of light and dark. A more current Western theory posits a disruption of electrical attraction between the protein and the bacterial cell that seeks it, a charged wall of sodium and chloride ions lined up between the hunter and its treasure.

Meat well preserved with salt will last a very long time. Preserved that well, though, it also becomes rather difficult to chew, outrageously salty in the mouth, as though the salt takes on the meat's strength in its efforts. In earlier times, when homes often had no water supply and much of the meat was salt-preserved, a cook was in a constant quandary. Most recipes encouraged the cook to soak salted meat in several baths of water, but if water was scarce other tricks were used: grains, dried beans, bread, potatoes, even pieces of linen could be added to a simmering piece of meat to draw out the salt. Mixed spices—rather incongruous mixtures of cinnamon, cloves, ginger, and more—were added, with sugar and fruit, to offset salt, perhaps by offering a distracting set of flavors impossible to ignore.

The origins of preserving food with salt are too distant to

discern. Whether the Egyptians learned first from embalming or from food preservation is unknown, but they became successful at both. And not unjustly; salt desiccates—makes pure by making barren—and dried food, like a dried body, is immortal in a sense. It is the organism in an elemental state— sun, salt, leather, bone.

I know this when I eat jerky—tearing the strips of shriveled and bloodless muscle apart, soaking each bite with my own saliva till it comes to a kind of life in my mouth, sere and salty, sweet and fleshy. The Peruvians used to make *charqui* by soaking bits of beef in brine and then rolling it into the cow's own hide for a day or two before both would hang in the sun for a final tapping out. *Charqui* became *jerky*, the mummification of meat like kings and pharaohs, till time for a new life.

Sailors knew salt pork and salt beef, and sometimes little else but hardtack and seawater and rum, for the long years of transatlantic voyage. Bored sailors amused themselves by carving pieces of salted meat into intricate snuffboxes, tributes to salt's long life. Sailors knew scurvy, too, and it was long blamed on salt. The money-conscious British Admiralty was happy to blame the obvious, about which nothing could be done. It was more than 200 years after the relation of vitamin C to the disease was discovered before they provided the expensive, irreplaceable citrus fruit required.

The interaction of salt and beef—salt and food—is a strange and tangled one. Salt denatures protein, evolves it, by breaking the bonds inside the protein molecule that keep each molecule discrete. Denatured, they cling to each other, making a larger, smoother lattice—making meat more tender to the tongue. Salt pulls water out of bacterial cells as well as the muscle fibers of the meat itself, and keeps the meat unadulterated. The loss of water increases the proportion of fat, so jerky has a smooth and oily texture. Corned beef is simply salted beef, named after the coarse salt called corn in times past.

Then there's nitrite. Nitrite is a salt, but of an acid. It

occurred as a trace contaminant in old curing salts, and by chance lends a bright pink-red color to cured meat that lingers even after cooking. Nitrites add a flavor of their own, too, a sharpness that many find desirable, and—as if that weren't enough—it slows the breakdown of fat.

Traditional methods of curing meat were 10 to 50 times higher in nitrites than modern meat cures, but then nitrites have been found to be carcinogenic, so it's probably for the best. Without nitrites, preserved meat looks less meaty, tastes less sweet and fleshy, loses a dose of its fatty oil. In a way, nitrites counteract salt's hearty efforts, the dry thinning of the cell membranes, the pulling out, the suspended animation of jerkied meat. Nitrites give them fresh, pink, false life.

Kosher is another kind of respite, an embalming in the name of God—to remove all blood. Meat is soaked and salted in a culinary egress, a perspiration, oozy and effusive, a drool of blood, evacuation; the salt sucks the fluid and expels it like a hemorrhage, draining, decanting, sprinkling the cutting board with lysed cells and liquid, till the flesh is chaste—the thing itself, faultless. Meat refined.

The Jews of old Italy made *carne secca*, salt-dried beef. A roast is covered in coarse salt, the kosher salt that sticks jagged to the meat to do its work, and weighted for a week while it drains. Then the meat is washed and hung to dry—for a month or more—hung in the crisp, cool air of late winter, a refreshing cold as winter disappears. Passover is to come before the roast is cut down and sliced for the Sabbath table.

Salt is a "corrective of foods," wrote the alchemist Paracelsus. "So many virtues be hid in the use of salt."

Yes, virtue: in cheese, bread, sauces, eggs, the boiling of water. Salt, to Paracelsus, was "the conservative element which prevents the body born with it from decaying. . . . Salt must be supplied to all."

First, to cheese, "milk's leap to immortality," as the kitchen wisdom goes. All but the very softest cheeses are salted. Some, like cheddar, are salted as curds, while others are rubbed on the outside with salt, like Parmesan with its hard rind to hold the center fresh. Swiss cheese is dipped as a wheel in brine, and the salt seeps in over time. As in meat, the protein in the milk denatures, allowing the protein whey to drift out where it can be removed (usually by pressure) so the cheese can harden. Salt slows bacteria to allow a slow ripening of a cheese, a liquamenish control. Feta cheese, the leap of a goat to heaven, is salted in a kind of passionate, impulsive, headlong flurry, salt as improvisation, the cheesemaker's artless and incautious fling.

Salt slows the yeast action in bread, again controlling its blossom to make for a ripe and fluffy dough. Too much can denature the protein bonds of gluten, toughening its texture, but in the perfect amount it prevents the same gluten from being destroyed by enzymes that would make the gluten sticky. It even colors the crust. In flour-based sauces salt is a mistake, its composite ions competing for water, making lumps. But the chloride itself prevents the discoloring of fruits and vegetables if one can stand the salty taste. An ounce of salt raises the boiling point of one quart of water by a single degree Centigrade—but then so does sugar. Any solid dissolved in water interferes in its boil, since the solid molecules don't turn to vapor at the same low temperature; they get in the way, take up room and heat. But in mineral-hard water salt soothes, bumping out calcium and magnesium traces that affect soaps and leave metallic tastes.

Cooking omelets can be done with salt; its denaturing work makes the eggy molecules pull together tightly. But in cooking scrambled eggs, where what is wanted is the soft, fluffy fullness, the airy puff, salt should wait till the table, else it flattens the eggs into a sheet.

As salt raises a boiling point, it lowers the point of freezing. From this is born the sublime, the real, the ice cream. Ice cream is the product of two unnatural additions, the sugar in the cream and the rock salt in the ice. Each lowers the freezing point of the respective fluid. Pure ice can't freeze ice cream because the sugar in the custard has already lowered its freezing point—from 32° to about 27° Fahrenheit. One part salt to eight parts ice is plenty to bring the cold brine below that point. (Except for sherberts and fruit ices, which have a higher sugar content and thus a lower freezing point.)

The crank turns, the icy brine pulls heat from the custard and starts to drop its temperature. That heat melts a little ice, and so does the salt, and the brine gets colder because salt water doesn't freeze. More heat comes from the custard and *it* gets colder, a part of it growing solid—but what is left as liquid has more sugar in it, and a lower freezing point still. It is theoretically impossible to freeze every molecule of an ice cream mix, and I know from practice that the ice never fully melts. The coarse rock salt clings to the cubes and holds them firm, melting their faces together the way salt cube faces melt in a rising dome, and slowly the brine gets colder and slowly the custard gets colder, and then the paddles themselves freeze up, and it's time.

If a boiling egg cracks, salt poured on the seam will seal it. An egg dropped on the floor will coagulate with salt, and pick up in a piece. Butter stays firm when wrapped in a cloth wrung with salt water. A hot pudding cools quicker set in a basin of salt water. Fresh milk stays sweet longer with a pinch of salt in it. A salt line stops ants, and salt water kills poison ivy. Salt is astringent, drying—brush your teeth with it and brighten the gums, massage a friend's feet in salt water and a rock-salt scrub till they're burnished pink—a "salt glow"— wash your hands with salt to take away the smell of gasoline, the stink of onions and sweat from clothes. Salt and lemon juice clean brass. There are good scientific reasons for this, clear-cut chemistry and ionic rationale. But I like to think of salt's virtue, of what Paracelsus called "man's need and condition of compulsion" for salt. It rises and buries, flies and sits, dries and wakens. In macrobiotics salt balances, a heavy to a light, but salt alone is unhealthy. "If we would try to consume minerals in the form of sea salt (yang) itself," writes one scholar, "we would quickly become inflexible and tense, and even fanatic and insensitive." But balanced with its partners, with dark and sweet and hot, salt heals.

Sam Clemens knew salt's virtues. He wrote to his wife: "I take only one meal a day just now and would keep this up if you permitted it. It consists of four boiled eggs and coffee. I stir in a *lot* of salt and then keep on dusting and stirring in black pepper till the eggs look dirty—then they're booming with fire and energy and you can taste them all the way down and even after they get there."

THE PHOENIX, the ancient symbol of treasure beyond compare, sits on a mound of dirt. A Chinese peasant stops in his labors, marveling at his good fortune in seeing the great bird mark this unremarkable spot. He gathers the dirt in a basket to take to the emperor as an offering, a gift he feels will ensure his future and the future of his son. The emperor, however, feels differently. Disgusted at the basket of dirt, he orders the startled peasant beheaded. A court attendant whisks the offending basket to a shelf in a dark pantry hall.

That night it rains. The emperor's servants scurry through the hall, bearing dishes of delight to the court. Rain leaks through the ceiling, soaks the dirty basket. The basket drips, drips into the bowls of soup and spiced noodles passing below. The emperor is charmed. What food! He sends all on a search for the magic ingredient. A servant sees the dull basket, now frosted in a fragile white, pungent crystals powdering its surface. Salt. The emperor finds the pile of dirt described by the unlucky peasant, grows rich selling salt to Asia, and honors his dead serf by ensuring the future of the son.

NERUDA CALLS underground salt "a mountain of buried light," and so it seems. While salt can be veined like coal and multicolored, it is often dug out of whiteness, shimmering in the miner's lamps, a thousand and more feet down. Here is the nest of the phoenix, radiance covered with dirt.

Salt is mined much like coal, but its shafts are more stabilized, safer, bigger, brighter. It is taken out "room and pillar," whole spaces cleared with support beams of compressed salt left behind. The ceilings may be a dozen or a hundred feet high. The hand chisel has given way to explosives and bulldozers and front-end loaders, blasting out the ancient halite in the hundreds of tons for sorting and sifting and the slow climb to daylight. The miners career through the cavernous white hallways in pickup trucks, throwing up small clouds of salt dust from the wheels.

Room-and-pillar mining, at least in the United States, is disappearing. Many domes are worked as brine these days, with water pipes pumping into the bowels of the salt and dissolving it into gallons of solution that can be pumped out easily, free of human hands. I find such a process utterly unromantic, even boring. If a few men dug a hole and poured hot water in by hand, listening for the gush and hiss of it hitting the buried salt, and then lowered a pipe and sweated on the handle of a stubborn pump—well, then, I could find a few things to like in brine mining. But the hallways—50 feet wide, high-ceilinged, rough-walled and all white, the light buffered by a fine salt mist thrown up by the miners' motions and the corners disappearing into gray darkness—that's a fine thing. The domes are so big, such tunneling is a mere ant farm in the earth. (Kansas lies over five trillion tons of salt; a few miles of mine shafts are bare pinpricks.) It seems almost profane to dissolve the hard-won strength and compaction of the millennia-old domes with mechanical pumps. Salt should be seen, and admired, first; a brine pump takes advantage of its chameleon nature, without apology or praise. Hallelu.

The salt thus mined, the millions of tons of it, is the top raw material used in American industry; it surpasses sulfur and coal and petroleum. Less than 5 percent is eaten—in spite of my help—and only another 8 percent is fed to livestock.

Water conditioning and chemical uses take a big share. But the biggest chunk—about 45 percent of all the salt used in the U.S.—is spilled on highways, to melt snow. The empty rooms left behind, warm in the depths from the earth's own warmth, move because salt domes move, empty bellies and all. But they are safe enough to store precious records, jewelry and food and microfilm and the original cut of *Gone with the Wind*, Bibles and wedding dresses and seed stock, coin and stamp collections and paintings. The strength of the dome, the steady temperature, dry air make a fine safe, hundreds of acres in size.

The salt mine in Wieliczka, Poland, has been in constant operation for more than 1,000 years. It is the oldest working salt mine, and holds in its depths a sunless sanatorium for people with respiratory problems. Salt miners everywhere have seen their mineral as marble, and the Poles are no exception; they've carved the rough-shod walls into sculpture, a mirror of the world hidden above.

A geologist visiting the Wieliczka mine in 1858 reported "chandeliers hung from the roof, with pendants of salt crystals instead of glass, and the walls were, I think, decorated with sculptures in the same glistening material," at 1,000 feet in depth. In the mine is a small cathedral, with a relief sculpture of Madonna and child carved from a wall near the altar of salt. One room is a memorial to Copernicus, another tells the tale of a ghost called the "Treasurer," who guards the earth, and yet another a 13th-century queen. Wieliczka began as a spring, retreated, and was dug as a mine. Now Wieliczka is a brine operation. Water floods new shafts distant from the respiratory sanatorium, and the galleries of shiny statuary seen by tourists.

Some mines aren't white; the salt may be twisted with silt and sediment into grays and blacks and stippled maculae. There is a large mine near Bogotá which invites visitors down a long, soft corridor of marbled, white-swirled walls to a huge cathedral. The ceiling disappears into blackness. At one side is an altar with steps and a kneeling board, and in front of the kneeling board a row of tiny pillars of salt on which the wafer of Communion is laid. Stalactites of salt like frosty icicles hang from the walls, and all is dim, large, still.

In one corner is an echo chamber, the floor carved in an intricate, repeated mosaic of color like parquet from different colors of salt, walls like black-green marble with thin twirls of white, a crystal chandelier above. Under the chandelier you can hear a voice, and feel the vibrations of sound shake your body. At the edge of the chamber, pressed against the salty walls, you hear only echoes, a faint sibilation of sound.

COMMERCIAL food-grade salt is not just salt, of course. Salt cakes without help—"when it rains, it pours"—either in its manufacture or packaging. Salt can be crystallized in certain shapes by heating the brine in certain ways; these imperfect shapes don't fuse together, and the salt pours. But often something else is needed, an additional desiccant to pull out the moisture that makes the salt stick: calcium silicate, magnesium carbonate, or magnesium chloride. Sugar is added, too, as a desiccant, and to stabilize the iodide.

Iodide is added to most commercial salts at 0.01 percent. Once it's added, more additives are needed to stabilize it, for the iodide ion is volatile and wants escape; its unpleasant, chlorinelike odor is an unfortunate side effect. The traditional stabilizer was sodium thiosulfate, now often replaced by dextrose. (One writer comments that dextrose sounds more innocuous on a package label.)

But why iodine at all? Thyroids. In 1907 research showed for the first time that iodine was essential to thyroid function. It took another 13 years to prove conclusively the connection between endemic goiter and iodine. A goiter is a massive enlargement of the thyroid gland, a globular tumor swelling out under the skin of the neck like a gnome attached to the throat. Tremors, lethargy, vomiting, anemia, disrupted metabolism all accompany goiters. Cretinism is worse, a condition caused by long-term lack of function in the thyroid, ruining bones, teeth, the heart, the brain. Endemic goiter is the result of iodine deprivation, and the difference between "goiter districts" and areas free of goiter is simply the amount of iodine in the soil and water. In India, 40 million people suffer goiters; in China there are 78 million victims. Neonatal hypothyroidism, or a congenital lack of thyroid hormone, is one of the two most common causes of retardation in the world.

The proof of the cause of goiter created an iodine fad, with people wearing bottles of iodine gas on strings around their necks, others eating so much in tablet form that they developed iodine-induced thyroid tumors. This paradoxical overdose, in turn, made iodine use a bigger controversy than fluoridation of water supplies, because goiter districts became the scene of massive public campaigns to increase the use of iodine.

In 1924, nearly two-thirds of the school children in the Lake Superior school districts had goiters. (The water in the Mississippi Valley has 10,000 to 18,000 times as much natural iodine as that near Lake Superior.) At this time every county in Michigan had goiters. The state banded with the Wholesale Grocers Association and the Salt Manufacturers Association to put an iodized salt on the market—cheap, easy, universal— and to educate the public. (Morton Company made the first commercial iodized salt, at a time when grocers were happy to encourage people to buy only iodized salt. Not bad for business.) Rochester, New York, soon added iodine to the water supply. Now most countries require the use of iodine in water, but not the United States—partly because iodized salt is so readily available. In the four years between 1924 and 1928, the goiters so prevalent in Michigan had almost completely disappeared.

We all need thyroid hormone all the time, but our need increases in times of stress, such as in the fetal period, during pregnancy, during the growth spurts of puberty. After iodized salt was made available, one Michigan county showed a drop in goiter incidence from 41.6 percent to 8.8 percent, and only one of the children who had used the iodized salt had a goiter. He was a 17-year-old boy who had grown almost a foot in less than a year. An author at the time noted the body's need for increased thyroid in puberty, and wondered how one could meet such "emergencies."

Iodine and ions, tiny things with a big bounce, small harmonic motions on the atomic scale, beating time. Michael Faraday coined the word for ion, from the Greek for "going," and so they do: slipping through the bonds of protein, muscle, and bone, tickling ice down and sliding enzymes to sleep, respiratory thrills of change. It's salt's pendular nature, a pea under the mattress, unseen but perturbing, noticed far more in its absence than its ideal dose. It is a saltatory dance, this dance.

11

CREATURES

CARNIVORES don't require salt in their diets; they eat the omnivores and herbivores that do. Carnivores literally eat the salt-filled cells and drink the salty fluids of animals that seek salt. It is a secondhand source, but convenient.

Herbivores need salt, and sodium in particular, because their diets are high in potassium, and naturally low in sodium. Without extra salt, milk production and fertility decrease, development slows. Animals know this need, in their viscera; a beef cow will eat 1 to 2½ pounds of free salt every month, a lick at a time.

Humans eat meat and crave salt, too, because for a long spell we've had grains, fruits, and vegetables in our diet. Meat is, in reality, a rare and undependable commodity for most people. The real puzzle is why some humans need less salt than others even on similar diets, and why some vegetarians don't crave salt at all.

The seeking of salt is a steady and unstoppable force in the animal world. A great basin in Illinois is worn several feet down in the center, the product of the hooves of mastodon, herds of which came to lick the salt that trickles from a spring in the center. Salty urine has long been an effective trap for migrating animals; salt licks in turn are a place of predation for the carnivores who don't need them. The desert bighorn sheep has been known to cross deadly plains, losing its young to predators and starvation, to reach salt. My cats curl under my arm, sleepy, and now and then try to sneak a lick across my shirtsleeves, under my arm, languorous addicts both.

All mammalian species show some appetite for salt, but as with humans the form has some importance. Dogs and cats are indifferent to a salty water at the concentration rats seem to crave. But a number of species—rabbits, deer, kangaroos, rats, calves—show a specific hunger for sodium, and the ability to distinguish it. The chemistry of the adrenal glands, specifically angiotensin, is as effective in disrupting the salt appetite in rats as it is in humans. Adrenalectomy, or removal of the adrenals (in rats), causes a steady excretion of sodium-rich urine, and the rat's hunger for sodium rises until it is willing to drink the most unpalatable and cloudy fluids, unable to satiate. Likewise, a steady infusion of angiotensin II into a rat's brains causes a tremendous increase in the rat's desire for sodium chloride and water—but not for potassium chloride. The rat distinguishes the two salts in solution, while its water intake exceeds its own body weight—until it dies.

A rat that has been deprived of sodium goes through a number of changes. The excretion of sodium in the urine drops dramatically and immediately. The amount of sodium in saliva drops, too. After a spell the sodium in bones is mobilized into the circulation. Given a choice, such a rat will choose to ingest only sodium chloride—not water, quinine, potassium chloride, sucrose, or any of a number of other minerals or solutions. The rat shows a hunger specifically for sodium from the beginning of its deprivation. If such a hungry rat has found a source of sodium in the past—even once—it will return to it when deprived, and overdo it. A salt-hungry rat eats too much salt and, researchers reluctantly admit, appears to enjoy its salt more. But all this requires a mouth and tongue; rats whose taste systems have been destroyed (there are several methods with which to achieve this) are unable to seek and ingest the needed salt, and die.

Rabbits and sheep, deprived of salt and then offered it, will eat only as much as their body requires. Rhesus monkeys, on the other hand, don't show any sign of salt hunger when it is simply removed from the diet. They must be "washed out" physiologically, rinsed clean of salt by diuretics that flush the blood and deprive the cells. Then, truly hungry, a rhesus will show salt hunger, and a bit of discrimination for sodium. In contrast, if a salt-deprived rat is fed salt solution through a tube into the stomach, given no taste of the crystal, the salt hunger remains long afterward. Tube feeding requires many long hours before the tongue stops longing for the salt, and the appetite fades.

Animals learn preferences like human children, and aversions from a single unpleasant experience. Rats will learn to avoid a salt solution if they associate it with pain. But they will only avoid that particular concentration of salt; solutions of higher and lower salt content don't scare them at all. Rats learn preferences from pain, as well—the pain of sodium deprivation. A rat that has been deprived of salt will ever after show an increased desire for it. Not a bad thing, really, considering the inexplicable nature of lab life. The environment is an unpredictable thing, and a smart rat stocks up for the future.

We humans lick salt off, and out of, each other's bodies in our foraging, a lover's skin as permeable as the unborn's. Mating and pregnancy are the mammalian climax, a physiological and racial necessity—the point. During pregnancy the appetite for sodium is most avid, even increasing a little during lactation. Take the rabbit. Even when her embryos are minute in size, microscopic, the rabbit's salt intake begins to climb. Her intake is two times normal in the first half of pregnancy, three to five times normal in the second half. She stores the excess in the fetus and placental tissues. The whole while she is producing milk for her young her intake stays four times higher than normal, and half the sodium content of her extracellular fluid is renewed every day. This rise occurs even if the litter is removed at birth, without ever suckling, the victims of the curious scientist in charge.

THE HUNGER for salt, and salt's exhalation. Seabirds have a salt gland; in most it is under the skin between the eyes, a smallish thing that empties into the nose. It is very light, and simple in a way—it pours out salt, a brine recycled, saltier than sea itself. Petrels eject it in a flying stream as they plunge with the air currents, a squirt of sea transfigured, and returned.

It is a puzzle still, a good-natured argument, whether seabirds drink the ocean. How do they get their water? Take it in with food in the way that we breathe air as we eat, accidental? Or do they rely on the fish they eat, soggy, waterlogged creatures that they are? One way or the other birds of the sea must rid themselves of salt. Pelicans, cormorants, the albatross, gulls—they depend on their salt glands. Without that gland all is lost.

Some birds, like the albatross, can't drink water sweeter than the ocean. If they do, all the body's salt dribbles out their nose, and they grow sodium-deficient and die. In the laboratory the albatross seems curiously ignorant, given a choice between salt water and fresh. It chooses first one, then the other, and its choice seems unrelated to physiological need. The salt-hungry albatross will drink fresh water, the salt-loaded one seawater, and vice versa. Why? Can't they, the way rats and people can, taste the salt? (Or is it the way I breathe oxygen, without flavor, noticing it only in its absence?) Or has the albatross never needed to know the difference? He is a pelagic bird, after all, salt in his breath and on the wind. The very notion that he might someday need to know the quality of the ubiquitous salt is unthinkable.

Flamingos have salt glands, too. So do ducks, eagles, hawks—and ostriches, and roadrunners. Just in case, perhaps. History, after all, is filled with earthly vagaries, geologic fickleness—ice ages, droughts, sunken lakes and newborn mountains. A species has to watch itself, lest conditions change.

The desert iguana has a salt gland. So do lizards. The lizard, in fact, has a small depression in his nasal passage, and there the salty fluid puddles, humidifying the dry hot desert air as the lizard breathes. When the fluid evaporates, the salt crystallizes and gathers on the lizard's nose, a beard, grizzled and white. (Even Darwin noted, at Galapagos, that the marine iguana squirted fluid from the nose when afraid. The salt gland responds to stress; touching a bird can make its nose run with broth.)

Two sets of definitions, then more stories. Creatures which can tolerate only a small change in the salt concentration of their environment are called stenohalinic—*steno*, Greek for narrow, and *hal* of course for salt. Loosely read it means narrow-salt-loving, actually. Its companion is euryhalinic, for creatures that handle big changes in salt concentration. *Eury* is wide, of course, for wide-salt-loving. This is a definition of the animal's (reptile's, bird's, insect's) response to the external. There is another word for the regulation of the internal, the extracellular fluid. Creatures which follow the breeze of the environment, melding their extracellular salt to the world's, are called *poikilosmotic*. A word of mixed roots, that one: *poikil* from the Greek for variegated, and *osmos* for impulse. A variegated impulse, then, for the little invertebrates, the quiet ones pushing along the ocean floor or floating in the tides. Their higher mates, the vertebrates like you and me, are *homoiosmotic*, of the "same impulse," conformers to an internal regulation. One way or the other we homoiosmotics have to find a way to rid ourselves of salt, and there are many ways. Many freshwater fish excrete sodium and chloride ions from the gills; others use hormones to change the internal metabolism of salt as the environment changes. Salmon eggs don't develop in salt water, and young salmon simply wait in fresh water until special salt-secreting cells grow. Embryonic fish placed in specific, abnormal salt concentrations adapt irreversibly to it, form and function, and can never readapt to their ordinary environments. Pickled outcasts.

How to rid oneself of salt? The tears, for one, of turtles and crocodiles and poisonous sea snakes. Manatees and dugongs cry, mermaid babies near the shore. Whales drink seawater in their heady passage, and then cry greasy tears to shed the salt off their fragile eyes. And urine: dolphins, porpoises, the harbor seal all have kidneys that adjust almost instantly to fresh water or salt. Whales, limited to the sea, make urine saltier than their blood. The bottlenose porpoise, full-mouthed ocean puppy, has glands on his tongue that seem to excrete salt. So he breaks water, and smiles, and drools a salty spit as he flies. And like the camel, whales and other ocean mammals make their own metabolic fresh water, from burning fat and oxygen.

One small word for the pygmy devilfish (cousin to the manta ray), which leaps 15 feet clear of the water for no apparent reason and slaps the surface with a resounding clap, uttering a musical bark as he falls. The dinner-plate, white-bellied, pop-eyed ray, furry to the touch as he slides past my hand, has blood a bit *less* salty than my own. Is he an old land creature, this delta wing of the smallish brain, with ancestral feet tucked under his chromosomes? And where does he put the salt?

Other species of fish—trout, mackerel, and herring, for example—keep their blood less salty than the ocean. They swallow ocean, leach out oxygen, and push the sodium and its fellows right back out the way they came, by the gills. These fish concentrate their urine, and can move from ocean to fresh water and back with little effort.

Then there is the shark. First of all, just to get it out of the way, the shark excretes salt rectally, through its rear. But the shark has another trick, a real piece of magic. It concentrates large amounts of urea in its body fluids—urea, urine waste, body poison—to keep its blood an equal thickness to ocean water, to prevent the osmotic leaching of differing concentrations. The urea is created in the shark's own tissue, held conservatively by the kidneys, wasted when abundant by the gills. Sharks commonly have a particular parasite, a tapeworm that lives in the shark's gut, a peculiar little string of unthinking worm adapted to the most unique of environments. And what does this little worm use for nourishment? Why, urea, of course.

But oh, the eel. He—or she, so hard to tell—is quite the study. The life cycle of *Anguilla* in its many forms is most capricious. The adults migrate from fresh water out into the Sargasso Sea, that independent pool in the upper Atlantic, and produce larvae which drift about on the currents for a few years, all stenohaline. Eventually they metamorphose into little eels, called elvers, and work their way back to the continental shelf, into estuaries, and then to lakes and rivers. These are unlikely shapes and sizes, too, weight doubling and then halving, first primitive, then less so, at one with tides and then fighting against them. And all along the way the eel's different versions of itself regulate salt in different ways—a new method for each stage, as though it weren't an independent being at all, but a form of the environment. The ocean's fetus, umbilically attached.

Well, these are small truths, hard-won, products of the laboratory as much as the shore. The odd lab rat, long bred for its defect and doomed to decapitation or the incinerator, teaches us a piece of mammal history, salt as a dry water that wets the cell as rain wets the ground. For me to know about salt glands—and I'm glad to know—a sacrifice is made. A duck is restrained, and a tube passed through its nostrils and out, and the salt gland fluid slowly sucked out by vacuum. Another bird is reamed from the rear, surgery into the bladder and rectum which, with little concern for scientific inquiry, empty into the same passage in birds, confusing the issue. Tubes must be inserted, sewn in place, then the necessary separation of fluid must be begun. I have a photo of a goose, its wings wrapped tight, head banded still, a tube of salt water pouring down its throat. I read of an experiment in cross-circulation, two ducks in tandem, their veins and arteries crossed and given each to the other.

The ducks sit quiet as decoys here, emptied and unruffled, deafened to themselves. They are strapped in steel bowls, weighed with care, clipped with skill. Many are decerebrate, brain-stem crushed and dry. It is altogether too much salt, an alchemical overdose pulling the life out, turning the self in on itself, turning memory to regret, and experience to grief. Too much salt, and some spills.

I wonder if it makes for a true illustration, all these artificial derivations, these artifacts of research like static ruffling the screen. A camel at the oasis is a truer picture, riper somehow. A petrel diving, a manta leaping in its own unspoken purpose. I sweat, and that's illumination, wiping it off my neck as I bend in the garden, salt on my lip as I lick it clean. The lizard turns his head to me as I pass by, dry in the heat, and bending close I see the flecks of powder on his chin, the hair of salt on the tip of his lizardy nose.

PART III
MAGIC

12

RELIGION

Sᴀʟᴛ Oʟᴅ Woman, the Salt Mother, wanders the Southwestern desert. She is a spirit of the Pueblo, precious, dangerous, and requiring great respect. "We hope you will come many times," say the people, when they address her. In the fall she is brought in, and her "veins are cleaned" in ceremony. It is a delicate undertaking. Anyone near with bad thoughts might turn into an animal.

The Amerindians had a curious relationship with salt. Some tribes ate salt, some never touched the stuff; those who did set limits. "A thing must be both fairly obtainable and fairly desirable before there would ordinarily be much motivation toward forbidding it," wrote Alfred Kroeber, the anthropologist. He was interested in the Indians, and their taboos; he studied salt in particular.

Kroeber made a map of salt use in North America, and across it wove a sinuous line: twisting along the Columbia River, following the Cascades south, cutting in a jagged edge across northeastern California to Nevada and then straight south, the line delineated salt. North and east of the line, natives added no salt to their diet, seemed not to care, had other concerns, other taboos. South and west of the line, salt mattered—and sometimes very much. Kroeber looked at diet, geography, climate. Nothing seemed to make the difference. In the end, Kroeber concluded, "the specific determinant of salt use or nonuse in most instances is social custom, in other words culture."

Culture, that most irritating thing. So hard to wrestle down, and name. Some herbivorous humans crave salt, some don't. Many societies introduced to salt accept it—learn to crave it, as it were—and some don't. A few cultures have traded salt, transported it across inhospitable deserts to profit from the cravings of others—and never taken a bite.

Taboos rise out of certain concerns: to keep the sacred sacred, to create awe. They are a form of sympathetic magic, that a person's behavior toward one thing can affect his relationship with something else. The limitation of pleasurable experiences also serves both to punish and to educate. Many taboos are specifically directed toward women and children, whether for reasons of bias or for protection depends on your reading of the culture involved. An anthropologist named Neumann has a particular theory about Amerindian salt taboos—that it is specifically a biocultural one, a response to observed physical changes related to diet.

Neumann studied the literature, most of it old and limited to the unschooled observations of travelers, and found descriptions of salt restriction in 11 tribes, including the Creek, Navajo, Caddo, Cherokee, Natchez, Choctaw, and Chickasaw. Ten proscribed salt in infancy, 5 proscribed pubertal boys, and 8 restricted pubertal girls. Seven tribes prohibited menstruating women from using salt. Ten prevented pregnant women from eating meat, and 1 from eating salt. Two tribes named salt as a prohibition during mourning. Neumann's heavily contested theory holds that there are reasonable physiological bases for these restrictions—or at least, that observed behavior could be assumed to relate to salt need.

The Cherokee forbad salt use during pregnancy—"salt makes meat swell," they said, in spite of their experience with jerky. And in pregnancy swelling (edema) can be a very serious sign of trouble, and is sometimes relieved with the restriction of salt intake. The Cherokee believed illness was magical—and since salt had magical power, its use was a use of magic.

The Caddo, similarly, believed salt made a person susceptible to magic, weakened their earthly defenses somehow. Caddo witches were able to kill white men through their food because the men ate so much salt. The Aleuts, though, claim that they see so well because they *don't* eat salt (though, like many tribes living largely on fish, they take in a great deal of salt through diet).

In old superstitions, salt provides power against fairies and demons, against all harbingers of death. It provides a spiritual, inner strength. The devil can't eat salt (and "the feast that wanteth salt is only fit for devils"). In large measure salt, the solid and the rooted, provides protection from the intangible pain of fear. Thrown in a fire, it wards off danger; thrown *at* a person, it reveals anger; thrown over the shoulder, it calms the restless spirit. Remember da Vinci's painting? Judas Iscariot turned toward Jesus in a moment of tension; his elbow spun out from his side; he bumped the simple clay bowl of salt, and the powder spilled a white wave across the linen cloth. Judas had his head turned, pretending loyalty; he didn't even see his mistake.

THE BIBLE made its role quite clear: "A covenant of salt forever, before the Lord."

Salt is binding, and bound: contracted, and contractual. It makes duty. Salt is foundation, fundament, the sprite of eternity. God required salt of the Israelites, in trade, a barter rooted in the earth's necessity. God asked for purity, perpetuity, in all things: "You shall not let the salt of the covenant with your God be lacking from your cereal offering; with all your offerings you shall offer salt." In turn he forbad leaven and honey at certain times, because they brought decay. Salt stayed. Elisha threw salt in a bad well, and made it clean again.

It is salt's iconic meaning that has given it a special place in Jewish worship. In ancient Israel the temples had a salt chamber to store the salt used in sacrifice; Josephus said that Antiochus the Great gave 375 bushels of salt to the Jews for this reason. The Talmud says salt must be used in all sacrifices. Salt of Sodom, salt from the Dead Sea, was used as incense in the old temples. (Two kinds of incense were burned at sacrifice: a sweet one and a maladorous one, to represent human nature. Every aspect of the human condition was present at a sacrifice, the lovely and the ugly both.) Salt is, comments the *Encyclopedia Judaica*, "the most common and essential of all condiments."

The Jewish table is symbolic of the temple altar; therefore the table always has salt on it, and bread except during Passover. During Shabbat, the Jewish Sabbath from Friday sundown until nightfall on Saturday every week, salt is central. The kiddush prayer is said over a cup of wine, then a blessing of thanks given over the challah bread, which is then dipped in salt and eaten. The salt is a solemn reminder of sacrifice and loss.

Salt of Sodom, the original crystal of the Jews, is a potent and strong-flavored season. It is high in potassium chlorides, stinging and bitter, acrid if a grain accidentally falls in the eyes. Out of this came the rule that hands must be washed after the Shabbat meal.

Salt is also the religious symbol of death, barrenness, and the forbidden: "the whole land brimstone and salt, and a burntout waste, unsown," reads the catastrophic warning from Deuteronomy.

Salt is a quiet symbol of amenable divorce for Jewish women. The Talmud reports that Shamai, a leader of the rabbinical school, interpreted the laws of divorce to say that a man could only divorce his wife if she had committed adultery. Hillel, another rabbinical scholar, allowed that a man might divorce his wife simply if she "burned" the food, which could be either by overcooking or oversalting. The traditionally liberal Hillel seems to be strict and cruel here, but in fact he gave women an easy out. Shamai's ruling would brand any divorced woman as an adulteress. Hillel's judgment, says an Orthodox rabbi now, was "divorce for no reason," the beginning of the "irreconcilable differences" plea.

In the desiccating barrenness of salt the Jews find a solution to a puzzle of God's—the forbidden blood. "Ye shall not eat the blood of any flesh. . . . Whosoever eateth it shall be cut off." The stern and strong voice speaks from the deeps of Leviticus. Not a drop of blood is to be consumed, not the blood spot in a fertilized egg, not a drop of blood from your own cut lip. The laws of kashruth are many and complex, the subject of debate and change. But salt is central, for it draws out, dries up the wet and bloody flesh of animals, shrinking the cells and collapsing the capillaries, squeezing the muscle like a sponge until it empties.

Christians have their uses, too, for the dry permanence of salt. The Venezuelans tell a story about Saint Andrew, sent to get salt so the people could wash themselves. But Saint Andrew got drunk, and forgot his mission, and in the end came back with too little salt. The men took it, of course, and ever since the women have been washing themselves with water. Until Vatican II simplified many ceremonies, salt was used in Catholic baptism, a small taste on the lip of the child or adult come to be blessed. The custom began with the baptisms of adult converts from paganism, as a symbol of their preservation from evil. As the priest slips the crystal on the tongue of the new believer, he says, "Receive the salt of wisdom." In a mirror of this the old pagan Irish punished men who accepted Christianity by limiting them to a diet of salty food without water— salt as punishment, resolution, execution, salt as the hero's trial and the unadorned tribulation of the Christian's way.

The hero's trial had many forms. To prove himself, a man must lift a great weight, or run a great distance, be more clever than a dragon or wiser than a Sphinx, measure the ocean or stand perfectly still. (Women are conspicuously absent from these stories; I wonder if they're home, testing the soup for saltiness.) One trial given in several tales is that of staying awake, vigilant, alert, for days on end. More than one hero solves the problem with his knife: he cuts a slice in his thumb, and pours salt in the wound. Who could sleep after that?

Outside the organized symbolism of religion, in the darker alleys of superstition, salt offers protection. You can kill a witch with a bullet of salt. Or reveal her—witches can't stand to be above salt, it is too pure, and if you suspect one, slip a grain under her pillow. She can't lie down then, it burns, and she'll run from the room in pain.

Moors carry salt in the dark to keep from seeing ghosts. (The ghosts are still there; salt grants a blissful ignorance to the traveler.) Africans believe hot salt in the eyes can kill. Carrying salt keeps a child from becoming a murderer. Salt water and absence wash away love (which one works best, I wonder?).

An English burial custom involved placing a pewter plate of salt on the corpse's belly—but not to prevent decay, nor to invite God's pleasure. Rather, it's simply a handy weight to keep the abdomen from filling with air and making the coffin hard to close.

The thirteenth Dalai Lama died in 1933 and was buried sitting up in a bed of salt, rested, serene. It is an old Buddhist custom not to bury a master for seven days after his apparent death, in case his stillness and quiet is merely deep meditation. Whether the Dalai Lama slept or dreamed or lay in calm reflection, a few days after his consignment to the powdery mattress his head turned; in the direction he had chosen to face, the wandering priests found the fourteenth Dalai Lama, two years later, when he was two years old.

"The people take their salt for granted," complained the Warrior Twins of the Hopi Legend, dirty little boys with poor manners—and strong spirits. The twins were angry with the way the people ignored them, and they moved the salt beds away, a long way over a difficult route. "That will show them," grinned the twins. Ever since, the gathering of salt has been an arduous and troublesome task, requiring a warrior's courage and strength. Salt grows in the earth like fruit, bud and root and blossom, but it dissolves, too; salt is transient, and fades.

Mt. Sodom has been quarried for salt for thousands of years, since the Bronze Age, since a hundred and a hundred generations past. The labyrinthine and dangerous chimneys and caves of the salty Mt. Sodom, plowed by rare rains and geological catastrophes, still look down on lost Sodom.

In Sodom in its worst days the angels came to Lot's home, and begged his hospitality from the mob. The tongue-clucking tradition holds that Lot's wife was stingy with her hard-earned salt, and refused to share it with the guests. Another, kinder version says that in all her fluttering graciousness she ran to a neighbor's to borrow salt for bread, and thus revealed the angels' presence in her home. The city began to burn, sulfur and brimstone to fall, and the angels guided Lot's family from the city. But his wife couldn't keep from glancing back, over her shoulder, to watch; she was held in the past, bound. She was rooted in a rain of regret like an old woman I know who fears salt, who feels it creep up through the hairs on her arm like a migrating rash.

Salt in solution, in waiting, dances liquid and flowing, as free as the water suspending it. But when it crystallizes to a solid, salt has a will and a way of immutability, a message of time. This is its duality and its unity at once, the strength of decision and the strength of change. Lot's wife, unnamed hostess to the inscrutable, self-proclaimed angels of God, found it necessary to look once more at her history.

She disobeyed, Lot's wife, she disobeyed the will of the men and angels, for a last look. And there she stayed, and stays today; of all the people of Sodom she, sculpted on the hill in permanent watchfulness, was the only one to see the whole story unfold. While a few fled and the rest died she had the best seat in the house, high on the hill, the dark sky lit by fire and the silence enlarged by screams. She watched the fires die out, the embers fall; she watched the next season's rains wash off the soot and consequently etch furrows in her salty toes and tear-ridges on her salty cheeks; a hundred years later she saw the first desert flowers emerge over Sodom and now she is the only one who knows where Sodom lies.

13

ALCHEMY

THE MANY-FACED nature of salt gives it many virtues, and only some are chemical. Though common salt, the molecule sodium chloride, does its work on the biochemical and geological planes, that work reverberates in less tangible arenas: in the mythos of place, in the customs of death and birth, in explanations for the inexplicable, and in the sealing of relations. The science of salt necessarily becomes the culture of salt and the science of culture, as much as water, air, earth, and fire.

The historian Hugh Kearney describes three major traditions of scientific thought, models he considers akin to linguistic structures. These traditions, he believes, converse with each other across the centuries like foreigners in each other's lands—opening the realm of the possible, adding accent and inflection and point of view. Each "thought model" of science is the product of the same desire—to find a key that turns the engine of the world. Different languages, one idea.

The organic model of science is Aristotelian, the legacy of the perpetually studious scholar, the seeker after fact. It is the universe in a state of constant change, all things alive and in motion, and that motion a natural one. Aristotle's universe was one of organization, a place for everything, every single thing in its proper place. There is a God in this universe, a logical, reasonable God whose actions can be presumed to make sense.

To understand God is to know the universe; to catalog the universe is to see the face of God.

The magical tradition, in contrast, views the universe as art, or mystery, a divine surprise. Who can claim to know its workings? This is a universe of hidden meanings, a daily miracle, a place of indiscernible attributes and opaque secrets trailing clues. It is Neoplatonic, beyond the human scale—Hermes's universe. As an Aristotelian seeks purpose and the reasonable chain of cause and effect, so does a magical thinker presume the need for transformation. They want to raise themselves, expand themselves to the proper size, at which the huge and occult world can be comprehended.

There is a third language (and many smaller dialects) spoken. That is the mechanical, the Archimedean, a world of predictable and cyclic change. It's a comforting place, God as engineer and the universe as blueprint, comprehensible. This is Newton's world, the world of political solutions and brave promises, clear-cut solutions based on dependable rules. Mechanists seek only for patterns—as opposed to inspiration. They want measurements, equations, sums.

These three languages, each bearing with it a philosophy and outlook the way English and Chinese presuppose the outlook of native speakers, are as divergent as conversing foreigners. They have in the passing centuries borrowed from each other, a word here, an idea there, bringing us to today's confusing vocabulary of world.

The modern scientific language is an odd beast. On the surface it celebrates the victory of mechanism—skill surpassing mystery. The only reason for learning *why* is so that we may *do*, and once we can do something, well then, so we ought. But underneath is a restless and uncomfortable concern, a salty itch, twofold. One voice asks: Is it possible to understand all this, this mountain of detail and differentiation? Where is the whole, the integral completion? And another voice echoes: Is that all there is to know? Here I am describing biology and brain, evaporation and history, many small pieces, a few with locking sides. The dilemma of the mechanist is that he breaks the machine in prying it open, and can't put it back together again. The search for secrets of function destroys the search for secrets of form.

Everywhere I look it is the same: people working hard at breaking things up. We try to divide, to disjoin the world so it will be easier to live in, less intimidating, palm-size constituents small enough to pocket. What we have for our troubles is a world growing more magical every day, like a cube of salt steadily growing faces, a mountain of the unknown as hard and bright as a track-studded mountain of salt. I want to go back a while, to a darker time with fewer answers, and find the place where salt entered: the salt idea, salt as sum.

HE WAS BORN Philippus Aureolus Theophrastus Bombast von Hohenheim, in the year 1493, and became known as Paracelsus. He is variously described by contemporaries: uncouth, quarrelsome, and boorish, vain and drunken, brilliant, maudlin, and rude. He was called a pagan and a scholar, charlatan, sorcerer, and fool, and was a bit of each. He bore a misshapen heavy head deformed by rickets. Some time later Goethe read his works and turned him into Faustus.

Paracelsus was born in Switzerland to a physician father and a manic-depressive mother who killed herself when he was 9, throwing herself off a bridge in a fit. He left for college at 16, to Paris, and began a lifelong struggle with authority. Paracelsus was cursed with an irascible faith in himself. He took his name after a great Roman physician named Celsus still honored in medieval times. *Para*, of course, is Latin for beyond. After seven years at school, he left to wander Europe in the plague years.

The science of Paracelsus's time was strange and strangely unbalanced, full of mechanical success born from religious magic. Fermentation, metallurgy, textiles, and the manufacture of glass and tools all were advanced to a high degree, but

in an entirely empirical way. Neither scientists nor practition-
ers understood the processes at work. The world was considered to stand on two principles, more metaphysical than
chemical: sulfur, the spirit, that which burned; and mercury,
the mind, that which vaporized. Paracelsus's Europe was a
Europe of Titian and Michelangelo, Luther, and Copernicus,
and Raphael; it was a Europe of lunatic asylums, famine, and
the plague.

Health and illness were connected directly to God and sin.
What was of man was of God. The body was not thought to
be a mechanical thing, a set of processes or parts. Animation
and decay were magical qualities, controlled by forces outside
human control, mythical, supernatural as all of God is beyond
man.

Paracelsus's professors gave lectures on anatomy from a
podium while a lowly barber dissected the body below. Physicians often made diagnoses without seeing the patient, relying on written descriptions of symptoms, before plunging
into a pharmacopoeia of pearls, mace, alder bark, ivory, and
roses (or the "dirt pharmacy": eggshells, snake blood, and
excrement). Apothecaries had no standards for dosage, no titration. A medicine worked because of its quality, not its
volume.

There was no concept of gravity. Philosophers expounded
on love, explaining that water flowed downhill because it longed
to be reunited with the ocean. All motion was light and heavy,
seeking its natural level: the light stuff, like air, rose; the heavy,
such as earth, fell. There were many facets to any one thing:
the thing itself, its color, its cosmic twin, a thing's operations
and uses, its principle. All that happened on earth and to man
was reflected in the heavens: numbers, stones, metals, and
the planets were all interrelated, immutable, divine.

In this unfamiliar world lived the alchemists, learned men
inheriting a Hermetic tradition of sorcery and adding to it
Aristotle's legacy of logic. The alchemists sought blindly, having no idea where to begin, for the Philosopher's Stone, the

Elixir, which would do much more than turn base metal into gold. The Stone would reveal the divine combination, grant its discoverer eternal youth, and cure disease. Many claimed to find it only to have it stolen or lost; others claimed to have destroyed it in fear of its acute dangers.

The underlying philosophy of the alchemists was a search for the essence, that thing that is uncreated and eternal. Perhaps the essence was a part of God, perhaps a fellow of God, but either way the search for it was an uncertain sacrament, a journey "pagan in feeling," wrote Jung, and possibly heretical in its very nature. The Elixir threatened to occupy a space near Christ, to reveal a light and knowledge unrevealed since Eden. Perhaps they looked for gold in order to avoid the full force of the divine.

"Alchemy makes perfect that which was imperfect." Objects were believed to be endowed with an internal alchemy of their own; this is what turned a bitter, hard pear or a sour cherry into a soft, ripe fruit. Metals and planets were connected by their qualities—quality itself was a mystical thing—so that lead was related to Saturn, because Saturn is the slowest planet, and therefore the heaviest. Terms were confused and ambiguous, the language lacked agreement. Substances had virtues and flaws built into their characters; alchemy wanted to find that which made things different, and the same. Alchemy sought a return from the fragmented and corrupt physical world to the pure plane of "quintessence," life at the source of the inexplicable forces of nature. Because the mechanics of substances as common as eggs were not understood—how a chicken made such a thing, how its hard shell rose from its liquid interior, how it changed form in such a substantial way—the search for a quintessence was a matter of trial and error. One philosopher buried the yolks of thousands of eggs beneath horse dung for two years, and then distilled the liquid that leached from the pile, thinking he could find that which was *egg* in an egg. When he failed, he simply tried something else: olive oil, and seashells. The organic was, after all, only one form of life. Metal grew in the earth like fruit on a tree, slower but in a similar form, bud and root and blossom. If fruit could ripen, why not metal? What made lead any different from gold?

CHEMISTRY has a curiously emotive vocabulary, a behavioral language that attributes, if only in word choice, a degree of autonomy to the inanimate. Electrons, the restless particles that circle an atom's nucleus in precise spheres of influence and purchase, are granted this the most. It is the particular work of an electron to seek balance, the lowest possible energy state for its atom. To this end they are said to accept and to donate, to share and to transfer, to become conjugate pairs. The virtues of a material, its alchemical quintessence, are the qualities that make the thing itself. The presumption that a thing *wants* to become itself, is always in the process of becoming its own self, still holds true in science. The quintessence of an electron, it would seem, is that of a diplomat— compromise, equivocation, moderation achieved by the most extreme and dramatic conduct.

A molecule of sodium chloride consists of an atom of sodium and an atom of chlorine, called chloride in its ionic form. An atom, in turn, consists of a central core of particles, protons and neutrons, encircled by the shells of an electron's path and domain. The number of electrons determine the atom's charge. Add an electron, take one away, and the atom is no longer neutral, but potent—empowered. Such an atom is restless; it wants, and seeks, neutrality.

Sodium doesn't occur singly in the earth, but always in combination. Sodium is sodium because of its structure: one electron only in its outer shell, and eight in the shell beneath. Chlorine has seven electrons in its outer shell. They are quite the pair; one electron shifts and makes both outer shells eight, equal. They bind to each other, two active substances forming an inert one. Sodium and chlorine become sodium chloride, a single thing, pacific, relaxed—and a crystal appears.

So MUCH OF alchemy before Paracelsus was the perfection of distillation: cooking down, evaporation, burning, making vapors. The search for the source of life, nature, animation became a search for parts. Alchemical distillation was a chemical extravasation of the world's blood. Jung writes: "The purpose of distillation in alchemy was to extract the volatile substance, or spirit, from the impure body." It was a more than physical effort; it was a distillation of the alchemist himself, a psychological or psychic process, repeated again and again just like the distilling of solutions, to increase the purity. The anonymous author of the *Rosarium*, an alchemical text, said that the Stone would only be found "when the search lies heavy on the searcher."

Transformation, like all navigation, happens by degrees. Paracelsus, for all his transcendent originality, needed many years of experience and failure—both practical and theoretical failure—to expand the fundamental philosophy he studied. He began with time to see distillation in a new light. Instead of taking the things of nature and finding the parts—a kind of incorporative reductionism—Paracelsus wanted to turn the crude raw material of the world into a finished product. He compared alchemy to refining. You could wear nature's skin of fur or have it tailored into a beautiful coat; you could eat the wheat kernels or bake them into fresh bread, an exquisite leap of *process*. He saw the possibility of a sophisticated purity—a cosmopolitan simplicity of the nature of all things. The quintessence was not in the undoing, but in the completion.

"[Nature] brings nothing to light that is at once perfect in itself, but leaves it to be perfected by man," he wrote. "To understand the laws of nature we must love her." The idea of essence changed in Paracelsus's mind. The Stone became the Principle—the thing turned into idea, like fur into coat. He came to consider alcohol the essence of wine, and blue the essence both of a blue stone and of the sky. He called the soul in the body "chaos," a primitive and disorganized state of matter "like milk in a cow" that leaves as vapor at the time of death. From there he made a great leap of imagination, and called gas the soul of matter.

In Paracelsus the organic and the magical traditions combined. He believed in the pure and logical behavior of the world, in the repeated and repeatable qualities of substances, but believed as strongly that nothing could be discerned without a great leap of the self: without becoming more like God. He felt that the interaction of the world's parts, the impulsive and spontaneous and volatile cracklings of life, had to leave a trace of some kind, a remnant, silt. The search lay heavy upon him; it drove him. As time passed he felt the inherent weakness in a two-pronged universe of sulfur and mercury, both fast and hot—it was an easily toppled structure. Paracelsus saw the power and equilibrium of threes—he saw the necessary result of necessary change. He added salt to the universe.

To TALK ABOUT salt—or, more properly, salts—you must first define a salt's components, acid and base, two particular kinds of ions. Modern chemistry has a number of definitions of acids and bases. Each delineates a certain body of knowledge useful in certain circumstances. It is an interesting kind of reductionism; the actual behavior of acids and bases, ions and electrons, is so mutable depending on the context that no single definition—no single *border*—applies. A chemist deliberately draws a border of what is useful to know *and talk about* at a particular time and writes his or her terms accordingly.

One definition, the first modern discussion of acids and bases, is called the Arrhenius concept, another gift from the man who first described the buoyancy of salt domes. It restricts the definition of acids and bases to their behavior in the solvent water. A later theory, sometimes called the solvent-system concept, defines solvents according to their capability in forming acids and bases of particular kinds. A number of other, less restrictive theories are in use, based on electron acceptance and donation.

In a broad sense a salt is simply a product of an acid-base reaction. Salts are grouped in categories and have an enormous variety of characteristics. (And beautiful names, too: galena and fluorspar, apatite, borax, nitre, and more.) Many are poisonous, glittering devils; they have their uses. A few are edible and essential. Some salts are neutral (called "normal" salts), others acidic or basic to a degree. Salts, because they are the product of acids and bases, dissolve into acids and bases.

Sodium chloride is a salt of the metal sodium and the gas chlorine. Sodium is "a degenerated metal," writes the chemist Primo Levi. "It is neither rigid nor elastic; rather it is soft like wax; it is not shiny or, better, it is shiny only if preserved with maniacal care, since otherwise it reacts in a few instants with air, covering itself with an ugly rough rind." I would disagree with Levi's dismissal of sodium's beauty: confronted with a fist-size chunk, dripping with mineral oil that holds the deteriorating oxygen at bay, I find it lovely. Irregular, a muted gray, it is soft enough to slice with scissors. The slice reveals a core as bright a silver as platinum, which seems to fog over like a mirror with the observer's wet breath. I slice, the sheen grows shadowy; I slice another sliver off. It is more like hard cheese than wax to the touch; a gray powder comes off on my fingertip and turns to sodium hydroxide, reacting to the hydrogen in the air and in my skin. This burns, and must be washed off right away. Dropped in water, the sodium floats, bubbling quickly into sodium hydroxide and gaseous hydrogen: it is said to *decompose* water. Fired, it burns with a stink and an intense yellow flame.

Chlorine is a gas, green and deadly; it combines explosively with oxygen and hydrogen. Often salt's commercial value is its parts, and their recombinations. Sodium chloride is separated into its ions in solution, by electrical currents. First the salt is melted—hot, 800° C.—and the molten flow conducts a charge. The positive sodium grabs an electron from the current, and reduces to the degenerate sodium; the negative chloride passes over an electron to the hungry anode pole.

Alternatively, a solution of salt and water can be charged. The water splits into hydrogen gas and hydroxide, the sodium chloride splits, releasing the chlorine, which is piped away. The free sodium and hydroxide then meet in sodium hydroxide. The concentrated chlorine is added to water systems, disinfectants, fire extinguishers, used in dyes, explosives, plastics, and more. Its companion sodium goes into paint, pigments, and photographic chemicals, bleaches paper and fabric, refines ore.

The two meet in a sum of attraction and repulsion, metal and gas into crystal. And the crystal, in turn, has wholly different properties. Salt runs fluid at 300° C., melts at 800° C., boils at 1413° C. Salt crackles when heated, the crepitating sound of hair rubbed between the fingers, close to the ear. Salt doesn't freeze in a different form; it solidifies at the magic 800° C., the demarcation between solid and liquid. Under the right circumstances of sudden severe cold, salt water can freeze as itself, rather than separating into frozen fresh water and salt. If the air surrounding it is windy, dry, and cold enough, the water molecules can be taken off, volatile, leaving a form of frozen salt, a salt ice, but in structure it remains the same. Despite our intimacy with it, salt is of the world, geological, lithic, inorganic in the most literal sense of the word: a mineral, never a cell.

All rocks are beautiful; they are the ridges and furrows of the earth's skin, chemical flowers. All crystals are beautiful, both atomic and macroscopic, slicing, slipping, flat, jointed, notched like arrowheads and meeting in zigzag seams. Sodium chloride, like all crystals, cleaves with the right stress, and the angle of fall, the design and rotation of the faces as they peel apart is part of what makes it salt. Salt may be white or yellow, red or purple, but it remains salt: consistent, bitter, vernacular. Salt is dry and free of impulse, it is the crust, that which holds and encloses—sometimes body, sometimes soul. All this because of two ions, meeting in a resolution of difference.

The bonding of sodium and chloride in their happy argument of charge is a linked chain, yarn woven in a knitted cube. It exists in a mesh of electron shells which are nothing at all but the viable path of an electron's voyage; from this *possibility* and nothing more rises all the virtue of salt—and likewise, of water and chlorophyll and air. The potential of shape and motion and response *is* the universe. What salt does when it crystallizes is bring salt's potential to earth, like the silent and invisible force of photosynthesis bringing green. If Paracelsus had known of the atomic forces, he might have stopped the search. As much as anything, they are the Elixir.

"SULFUR IS that which burns, mercury is that which embodies the virtue, salt is that which holds the body together," wrote Paracelsus. Salt allows mercury and sulfur to work. Mercury is wood, sulfur flame, and salt the ash. Sulfur is gas, soul and energy, oil, fat, and resin—it represents the celestial. Mercury is fluid and intellect, vapor, liquid and thought; it lives in mathematics and space. Salt is willing and visible, shown as a square and called the moon.

Salt is mass and balm, tangible philosophy. It draws down, and out, like a kashruth of the psyche, a leaching out of the blood of the soul to find the matter refined. Each of these three *represents* its virtues the way I represent myself; the visible form is only one face thrust into the spectrum. Humans are microcosmic, containing all the parts of the universe in proper ratio: not only salt and other minerals, but metal, gas, stars and planets, gravity, mass, inertia, flame.

Salt "is the honorable prize in the gift of Nature," wrote one of Paracelsus's biographers. In his time Paracelsus saw it used in making glass, smelting metals, glazing pottery. He wrote long treatises on the medical uses of common salt. "God has driven and reduced man to such a pitch of necessity and want that he is unable in any way to live without salt," he wrote. "I have said of salt that it is the natural balsam of the living body. That is, so long as the body lives, so long the aforesaid salt is its balsam against putridity. . . . If salt can preserve the dead body or corpse, much more will it preserve the live flesh. If by its power and efficacy salt preserves the dead body from worms, much more the living body. . . ." He felt that man's desire for salt was the desire of the man's own salt for itself. Salt is the self-referencing element (and as Jamaicans would have it, "Salt never says it's salty"). Paracelsus thought urine was a form of blood, and called it "the salt of natural salt." He chastised physicians for failing to recognize its power, especially in the cure of the "tartaric" diseases, like arthritis and gout, diseases that resulted from an internal precipitation—a stinginess.

Paracelsus brought this notion of salt as medicine to the European mind. As its virtue grew, it ceased to be sodium chloride, and became alchemical salt instead: psyche, mind, the quintessence of humanness, a common and then more than common thing. Salt was the key that opens and closes, a clue to the world's animation, its immensity of flavor. They called salt Oroboros, a cardinal substance. The mineral was only a shadow thrown by Salt itself. Salt was the separation, drawing down, binding, the residue. No single molecules of sodium chloride in the crystal, and no slicing off of salt from Salt. Salt, in its solidity, was what Jung called the "individuating mind."

SALT REMAINS, and salt disappears. Throw the crystal in water and face by face the ionic planes peel away, each newly exposed face sliding off the shrinking cube. To understand solutions, which are part of salt the way the individual is part of the collective, requires a certain understanding of water.

Water is an impossible fluid. Water laughs at reductionists—taken in its component parts, seen as a sum of particulars, water should not be water, exactly. Similar, yes, with certain qualities, but not all the qualities of water, not the variety and surprise, the promiscuity of water.

Water is the only inorganic substance normally liquid in the earth environment, and the only natural substance found in all three states—solid, liquid, gas—under ordinary conditions. It has the greatest surface tension of fluids, it is expandable, compressible, solvent. One could say water is the world's solvent, what the world is made of alchemically, solvating incompatibles to new phases: the solution of tears, blood, urine, semen, the sweat of God.

Water molecules, two hydrogens to an oxygen, form clusters together in liquid, bound by strong hydrogen bonds. As the pressure on water increases the clusters break apart, freeing individual molecules—the water becomes more fluid, more *watery*. If the pressure increases further, the free molecules can't form new clusters and can't move as much, either; the water is less fluid, denser, more syrupy.

When an ionic substance like salt is thrown in water, each ion is surrounded by a sheath of water molecules (a process called hydration); this water cluster is less fluid, more dense, strictly oriented and limited in its motion. Gradually a measure of salt dissolves in water to form a solution of water and sodium and chloride, a balance of dissolution and recombination. Solutions are not mixtures, like oil and vinegar in a jar. They are phases, singular, new. Not salt, not water—but *saltwater*. The desire of the ions to reunite is matched with the desire of the water to surround. When an electrical neutrality and a neutrality of clustering is reached the solution is saturated. The water is soaked with salt, like a rag mopping up a spill.

(Salt "salts out" other solutes. If sodium chloride is added to a saturated solution of carbon dioxide in water, it bumps the CO_2 away from the water molecules—the water prefers the salt. Add a pinch of salt to a soft drink and watch the bubbling increase. The same sort of thing is done in soap manufacturing. At a certain point the soap itself is in solution and must be removed; to achieve this, salt is added, and the water seeks the salt and leaves the soap unattended.)

At room temperature 100 grams of water is saturated by 36 grams of salt. Stir in teaspoon after teaspoon until a fine sand of salt settles on the bottom of the glass; this is the excess salt that can't be dissolved; there is no more room. But . . .

Heat the saturated water, bring it to a boil. As it heats the water seems oily, alive, with shimmering shifting lines like the crooked rectangles and triangles of skin, rolling in and out of shape, remaking themselves. When it boils the sides of the pan coat with a white splatter like paint; a hot and almost burned smell rises. Distillation, of a kind—vapors. Throw a teaspoon more of salt in the boiling water, and the ions peel off in spite of saturation. Heated water, like pressurized water, has fewer clusters and more free molecules, eager to crowd. At 100° C. 100 grams of water will dissolve more than 38 grams of salt.

This water is supersaturated. It is more than full. It is a jar of power, balsam and ash in motion. Cool it carefully, see the skim of sharp-sided powder like unmelting snow. Be careful not to bump it, and it remains a solution at room temperature—ripe, parturient.

"STOP MAKING gold; instead find medicines," he instructed his colleagues. He ceremoniously burned the main textbook of medicine, the *Canon* of Avicenna, in a self-conscious imitation of Luther. "God did not choose to give us the medicines prepared," wrote Paracelsus. "He wants us to cook them ourselves. . . . Medicines were created by God. But He did not prepare them completely. They are hidden in the slag." This was no oversight or failure on the part of God. He intended man to work. Just as a thing is made fundamental when it is finished, so man becomes purely man and the child of God by doing God's work. Man is finished by finishing the world. Here Paracelsus began to change a two-pronged philosophy of science, the organic and the magical, into a balanced trio. He sought patterns and rules. He added the mechanistic to medicine.

In practice Paracelsus did little new, except for one change of the most profound effect: he created a system of dosages. Until Paracelsus neither pharmacists nor physicians spent much worry over doses; it was virtue that worked, and the purity of a substance's virtue was what mattered—not the amount. Paracelsus said that amount, titration, was the difference between poison and cure; there was the possibility of too much of a good thing, or too little of what's right. This made "virtue," "essence," a quantity, a measurable thing—it made virtue chemistry. For the first time the vaporous essence of the universe, the force, the key, the idea, had a physical presence. There was *something* that was virtue and it was not etheric. It was, so to speak, the arrangement of vacuum between the atomic particles that made a virtue virtuous.

He claimed miracles; a cure for epilepsy, for cancer, rabies, syphilis, and more. He wandered Europe, stopping in the homes of strangers, killed their livestock and burned their pots to make his potions, filled their kitchens with noxious fumes and sticky perfumed syrups and—so he said—made the lame walk and the dying well. His doctor's hands fell far short of his philosopher's mind, but made medicine modern anyway, almost by accident. Paracelsus created a vision of the body as a set of parts—microcosmic, perhaps, but still parts. And then he went another step, and said the body could be influenced in foreseeable ways, by substances in particular amounts. If sulfur is soul, then there is something soulful in sulfur, and something sulfurous in the soul; if mercury is intellect, then the intellect is mercurial, in a real and solid sense. It was a grandiose and utterly new idea: man as machine, and man as engineer. And to balance the triad, salt, a real and tangible rock, a powder with the miraculous power to halt decay and hold the body in suspension, giving off some invisible but absolutely present quality, philosophical and dry as earth.

Whether in utter solemnity or with tongue in cheek, I can't tell, but some alchemists took to calling salt "Lot's wife." Paracelsus, in all his psychic extremity, held fast to his own superstitious and magical tradition, giving a literal *bio*chemistry to humanity and retaining God for medicine. "When we administer a medicine," he wrote, "we administer the whole world."

Hard changes, for Europe and the physician both. Paracelsus was torn between natural knowledge and divine knowledge, the Elixir and Christ. "I also confess that I write like a pagan and yet am a Christian," he said. He named a portion of one of his books "Pagoya," a neologism of the Latin *paganum* with the Hebrew *goy*. Knowledge of disease, he never quite forgot, was pagan knowledge because it derived from nature. Jung felt that Paracelsus's search for medicine was a search of the conscience, that he suffered the pangs of the "pitiless judge" hidden in the heart of original sin. Paracelsus was salty: hard, demanding, bitter, strong. He couldn't quite integrate his multiform culture and the breadth of his own mind. He could never escape the unconscious fear that not only knowledge of the body, but knowledge itself, was wrong. Nor did he escape the need to know.

Europe gradually left the plague behind, and with it much of its organic thinking. The mechanistic tradition, embodied in Galileo, Descartes, and Pascal among others, proved too potent for the interpenetrating and opaque world of the alchemists. The circulation of the blood, the revolution of planets, behavior of gases, the law of gravity—the physical plane was indisputable, and near. Its laws were consistent, its pattern predictable. Alchemy was, as James Hillman writes, "soft-edged"; alchemy left room for doubt and wonder—room for God.

Not so very long after Paracelsus died, a man named Kozak added his thoughts to the meaning of salt. He considered it evil and violent, a source of decay and putrefaction as well as of generation. Kozak saw salt from the punitive perspective of religion: that illness was sin itself, and a punishment for sin. Disease was the damage done to the body by the soul's fear of retribution, and salt a symptom of that evil. A salty body was a body held in thrall from death and judgment.

Gradually the mystical side of science changed to such a fear of vitality and the unseen. What Paracelsus called the administration of "the whole world" has become mere molecules. Concern for the relationship of one thing to another, one creature to another in a universal web, became a concern for the differences. Pattern conquered form; rule conquered desire. But the world remains the same: the quintessential animation, the *egg* in an egg, the *gold* in gold—the *nature* of nature—hasn't changed. All that has changed is our view.

A salt solution left alone, slightly cloudy from the impurities, will gradually evaporate. Each water molecule lost to the air frees ions for each other. Crystals grow—tiny seed crystals which grow into giants, cuboid and firm. In evaporation, writes Hillman, "the steam, the smoke, and the cloudy vapors ascend and dissipate in hot air. We lose the lushness of feelings, the flush of high hopes, the dumb bogs of inertia; and as the moistures recede, something essential crystallizes in the dry air." I have a bowl of crystals in my kitchen, like a scrub forest under snow, a powdered desert; they rise sparkling like frost on grass, swelling on the string I stretched days ago in the now-vanished water. An intelligent simplicity, focused. This is why contracts have long been sealed with an exchange of salt. "Salt is between us." Out of the invisible is the manifest, moving in an ancient and perpetual motion. This is the necessity—and the desire—of the thing to become itself, and then be done.

14

CURE

THE THING itself. What makes it? (Wouldn't the maker of the thing be the thing itself, too?) Where to find the force that holds us together, pulls us apart, and tame it?

"Causes exist in such subtle form that they cannot be seen by eye," wrote James Tyler Kent, an American homeopathic physician. He was a leading practitioner of the healing system of homeopathy, a system both oddly illogical and undeniably obvious in approach. Homeopathy is more than 200 years old, practiced by hundreds of thousands of people around the world, heavily researched and explained in lucid detail in hundreds of books; homeopathy is also rejected as the worst kind of quackery by most allopathic physicians.

The Law of Similars is at the heart of homeopathy: "Like cures like." A substance that would cause certain symptoms in a healthy person can, if used properly, be the remedy for those same symptoms in a sick person; the unique force that makes a thing itself has the power to stimulate the body's own healing. In homeopathy the symptom is not the disease—the *disease* is not the disease. Illness is a disruption in the vital force of the person, that which can't be seen, in our Elixir. "The conviction that disease is caused by bacteria is probably one of our greatest illusions," writes George Vithoulkas, an internationally known homeopath. Health, he writes, is "not at all a question of killing bacteria but of bringing the whole human organism into a state where it is impossible for bacteria to thrive on it." Illness is never only physical, but a state of imbalance in all aspects, often seen first in the emotional or mental sphere: "To a homeopath, the patient is ill when and because he feels ill."

But homeopathy is not limited to the cure of bacterial disease, either. It cures the person, not the symptom, and with the cure the symptoms—arthritic pain, depression, sleeplessness, sore throat, the fear of love, grief, a desire to scream in public, a gouty toe or a tumor, a rash, conjunctivitis, nightmares, anorexia, or asthma—disappear. Symptoms are the body's response to turmoil, its best effort under the circumstances, a twisted but well-meant attempt to heal. Repressing symptoms —using antibiotics to wipe out bacteria, anti-inflammatory drugs to ease the swelling of arthritis—only makes things worse.

Homeopathy is "the medicine of an unknown science," writes Richard Grossinger, in a comment both on homeopathy's quirky philosophy and on its undeniable efficacy. Accepting homeopathy requires an acceptance of the idea that there are unseen forces at work, *nonphysical* forces at work, an idea that seems problematic when taken at face value. But considering the energetic nature of the atomic forces, the effect of placebo drugs, and the lights of the aurora borealis, the twist of starlight as it passes around the sun's bulk and the release of radioactivity in the splitting of a nucleus, considering the voices a schizophrenic hears from a fire hydrant and the compression of coal to a diffracting transparency of diamond, and considering the nature of dreams and the germination of a seed—accepting unseen forces is something we actually have some practice at doing. Homeopathy requires us to take the idea of immunology expressed in allergy treatments: a bit of bee venom, for instance, injected into the skin of a sensitive so as to induce the growth of antibodies against that venom for the future. Take that idea and unfold it to include the psyche and the heart, the personality and the character and hopes of the whole person. Inject a bit of the nonphysical force that ails us, like hair of the biting dog, and kick the body into balance, wake it up, shake the psyche by the shoulder, bring things in line, till the system stands up straight again. If a person presents a symptom that could be caused by a certain substance in a particular dose, then that person could be cured by the *force* of that same substance,

Thus, belladonna causes convulsions, when given as itself. Fleas cause prickling, itching sores on the skin. The solution to convulsions (of a particular kind) is the *potency* of belladonna; the right medicine for a certain itchy sore is the power of the flea. But it is neither belladonna nor flea that is given as cure. It is the force of belladonna, the revealed and radiant vitality of flea. It is the *belladonna* of belladonna that cures belladonna's pathology.

How to find this force, this power? The alchemists distilled and vaporized. Paracelsus sought the proper combinations that would make the single substance. Homeopaths dilute, and with each dilution the force increases. It is as though the freeing of a substance from the physical plane, as the amount of its mass decreases, permits the release of its nonphysical energies. Crush the flea (and various methods are required, care with one's medicine) and put it in solution, and shake: each shake is called a "succussion." Then take one-hundredth part of the solution and add it to another vial, and another one hundred succussions. And again, dilution, succussion. The process takes the powers that are "hidden and dormant" in the physical form and rouses them, said Samuel Hahnemann, an eighteenth-century physician who invented the theory and practice of homeopathy. "Somehow, the repeated dilutions and succussions of a homeopathic medicine release a great curative energy which is inherent in the substance," writes Vithoulkas. Alchemical distillation—the liberation of quintessence. The flea of flea, most wholly itself when its potential is exposed.

One needn't believe in homeopathy to be cured, either. The force is the same either way. I knew nothing of homeopathy when I found my way to a homeopath, suffering a chronic asthmalike cough, dry and hacking, after a rough course of pneumonia. I suffered a growing phobia at the same time, a kind of agoraphobia that made it increasingly difficult for me to speak to people or leave my house. I presented the world with a confident and even gregarious face, and was eaten inside by a terror of society. And I coughed, my hands and feet were cold, I had a long history of digestive complaints. For two hours I listed a series of problems, none related to the other in my mind, until I exhausted the litany of all that was wrong with my life. The homeopath, a young woman, told me to stop taking the anti-asthmatic medication I relied on several times a day, and instead to take one single dose of lycopodium, or club moss, bound in sweet white pellets. I did as she suggested, and the cough disappeared, never to return, within two days. Gradually the paralyzing fears disappeared, too, and my digestion improved, my tolerance for wintry cold improved.

I've since learned that lycopodium is a common remedy in the United States, often referred to as the "coward's remedy," for people like me presenting a brave and assertive face to a world that knows nothing of the quaking insecurity inside. I've returned to homeopaths for years, with much satisfaction. My remedies tend to be more common ones, types reflecting the peculiar personality of American bravado and fragility. Remedies like *Aurum metallicum*, the long-sought gold, with its tendency toward despondency, congestion, fear, a sensitivity to noise, and sleeplessness.

One of the most commonly used remedies of modern homeopathy is *Natrum muriaticum*. Salt. Table salt is "potentized" by dilution, and becomes a remedy of psychic salt—an alchemical vitriol. But what does salt cure? In old folk beliefs, plain salt cures bursitis, asthma, infertility, and the croup; it eases earache, gas, headache, pneumonia, snakebite, and warts. But that is salt—sodium chloride—with its crystalline virtues. In homeopathy, as in alchemy, a substance's virtues aren't bound to its chemistry, but to its character. Salt sinks. Salt remains. It dries, shrinks, empties. Salt evaporates, reveals, and is revealed. Salt holds—salt drinks—salt separates. And like all homeopathic "constitutions," as the complex of symptoms of any remedy are called, the constitution of salt reveals salt's nature in all the planes: physical, emotional, psychological, psychic.

Salt is the archetype of the individual in thrall: the single freeing itself from the collective, the ego in a state of separation. The experience of emancipation from the group is always a difficult one—it is the journey of psychoanalysis, the journey of religion, the child's journey. Many people find it too painful, thinking there will never be a road back, and remain in between—unable to return, unable to complete the passage. The many contradictions of what a homeopath would call the *Nat mur* type of patient—the type curable by a potentized salt—are the mirrored contradictions of salt. Think of them as the complementarities of the character; contradiction is the symptom.

The *Nat mur* person retains. He retains water, perspires either too little or only in certain circumstances, retains wastes in the form of acne and constipation. The *Nat mur* woman retains menstrual blood and fails to have a regular flow. Such a person can't urinate in front of others. Headaches and bodily pains are stubborn and unabated. He is unable to cry except when alone. He retains impressions, too, puzzling over hurts and worrying supposed slights for years. He can't forget, mourns for long periods of time, often appears taciturn, cheerless, isolated. *Nat mur* is a loner who is always lonely. At the same time that *Nat mur* seems to cultivate painful memories, he dislikes consolation; he longs for romantic love but only loves inaccessible or unattainable people. Loyal, almost tactlessly honest, reliable, steadfast, the *Nat mur* type is also reserved and stoic, reticent, hidden. He seeks approval but can't conform to social expectations, wants routine to the point of compulsion but may suddenly reverse an opinion. *Nat Mur* chooses the hard way and follows it with a stubborn tenaciousness; his love of truth is partly a product of rigidity. He suffers Addison's disease and goiter, and sometimes a longing for death that has no impulsiveness in it, no guilt—it is instead a tiredness of life, a sense of being worn down. Nothing fulfills.

Clumsiness, spilling, awkward movements and speech, a lack of balance—this is *Nat mur*. There is a deadness in it, weakness, exhaustion, anemia, the barrenness of plain salt. " 'You aren't going to *change* me with the remedies are you?' is the anxious inquiry," writes one homeopath in description of the *Natrum muriaticum* patient. Then the same patient will suddenly reverse his life, leave a marriage, a job, change all his opinions and claims. The crystal of salt moves into solution and a new phase, a new form; as suddenly it recombines, solid, apparently immutable.

The *Nat mur* patient is drawn to sun and sea, and suffers from both; he is "hurt by that which he loves best," writes Catherine Coulter. In the imbalance of the root of salt, the sources of life are in conflict with a disgust for life.

Salt tends toward precipitation; the phase of solution is only a temporary one, requiring a continual topping out, a proper ratio to be maintained. Salt holds the impulse in check, but too much kills impulse altogether. Our psychic salt is memory, the licking of wounds, lessons learned. The individuating mind of Jung becomes, with too much "salt," the self-referencing ego. Salt is the thing that seeks to know itself, to become itself. It is the purgative of experience. Overdone, it dries up the lifeblood and leaves the flesh an arid woody shell, neurotic and self-centered.

Wilhelm Reich said physical symptoms weren't the causes or effects of the psychic state but *were* the psychic state in the physical, somatic plane. The planes, or bodies, of a person don't interact in a strict sense, because they aren't separate from each other, but manifestations of a single thing. So salt is common salt, a psychic immutability, an emotional lability, a physical retentiveness and a somatic desiccation. It is a rooted substance and a root symbol. Salt is that which holds together, precipitates, stays and at the same time liberates. It is both cynicism and wisdom, duty and the fear of intimacy, the body in check, the mind at rest, the emotions in ease. Salt seasons the self so that the self's own true flavor emerges. It is above all a latticework that crosses and balances the planes: blood and sweat, tears and urine, the streams of sperm and ocean water, rock and bone and soil in slow, deep furrows from the cold depths up, the mind's leap of self and the sharp fate of the world's reaction.

PART IV
CHANGE

15

WATER

Any single ion may travel a million years, from water to rock to mountain and ocean and back. Any small cube of salt may dissolve, grow, disappear, crystallize, hide, reappear again and again. Such a giant wheel turning, such an unstoppable force, like slow cold furrows rolling out from the deep, sliding back to the antarctic cold, sinking once more. Salt spins round the same as water does, and in the earth salt and water are commingled as much as they are in the kidney, in sweat, in blood. Water, water everywhere, and with it, always, the salt.

There are great bodies of water lapping under the soil, splashing up against the sides of the earth's foundation, leading secret sea lives, tuned to secret tides. These are the aquifers, massive filters of sand and gravel and loose rock layered with water. The water is old: precipitate of rivers and lakes leaking through their floors, or connate water trapped molecule by molecule in sediment. These are fossil waters, the products of more humid times. (Some deep aquifers are new, filled with the juvenile water of magma, volcanic spit from far below.)

There are two kinds of aquifers, confined and unconfined. The confined type is entrapped between layers of rock less porous, like granite or shale, so that water slides off the hard surface on either side and sinks into the sandstone between. Confined aquifers are under pressure, and produce artesian wells bubbling up on their own like serendipitous, spontaneous, life-giving springs. Aquifers that lie horizontally, above and below less porous rock, are unconfined; their borders shift and are less definable. The groundwater is always in motion, like sodium, in and out, the levels dropping and topping off. Unconfined aquifers are quiet, without pressure; their water must be pumped.

Under the High Plains lies an unconfined aquifer called the Ogallala, made of sand and gravel, filled with water tens, hundreds of thousands, perhaps a million years old. The Ogallala is the largest separate aquifer in the world, covering 150,000 square miles and filled with enough water to top off Lake Huron. In places the Ogallala dribbles to the surface as lakes, rivers, and marshes, but for the most part it lies under arid, empty land—an area that, until a scant 100 years ago, had never sustained a permanent civilization. Now the Ogallala irrigates 10 million acres of farmland, but nothing irrigates the Ogallala. It is a natural charity, soon to disappear for good.

The steady pull of water from an aquifer is called "water mining," and it almost always means an "overdraft"—more is removed than replaced, and like all overdrafts this bears a charge. The precious gem of water is removed, one molecular jewel at a time. There is nothing subtle or slow about the process, either; overdraft in Arizona in the 1960s every *year* was equal to the annual runoff of every river in the entire state. In parts of New Mexico overdraft to natural renewal is equivalent to five feet of water out for every half-inch of water in. Billions of gallons of water disappear, and the Ogallala is rapidly running dry; not in our lifetime, nor in our geologic era, will it be replaced.

Before the Ogallala turns dry, though, it will likely turn to brine, a process already under way. There are several reasons for this, but chief is irrigation, the bringing of water to land that had no water. No matter if the irrigation is done by diversion of creeks or rivers or with water pumped from wells. Either way brings trouble: inland seas of salt water and empty flat salt pans, Deuteronomy's "whole land brimstone and salt."

The Ogallala, like all aquifers with an edge near a coastline, forms an underground transition zone, an area where the fresh water gradually grows saltier as it approaches the ocean from its rockbound bed. The pumping of water decreases what pressure there is, and the transition zone moves landward, crawling saltily into the rocks. Salt water intrudes on sweet, a natural response to the change in the pressure gradient, and the salt content goes up, and up, and up, tangier, tastier, deadlier.

Farmers like to irrigate by flooding, and thus irrigate less often. Doing so without perfect drainage causes the immediate water table to rise, and bring salt with it from the rocks below. Then the field evaporates and salt is left behind, damaging crops, encouraging the farmer to flood again in an attempt to flush out the salt. If the field is uneven, flood irrigation always leaves some portion high and dry—and the temptation is to flood to that lonely highest point. Irrigation actually makes the salt in the soil worse, because the water drains only to the point of impermeable clay, and then rises, into the root zone, full of salt and minerals. (Thousands of acres in the San Joaquin Valley of Calfornia are out of production now because of salt; the fertile soil rests on a bed of clay in some places only a few feet down; what fresh water there is on the surface evaporates swiftly in the hot, dry summer air.)

Aquifer mining causes a structural change called land subsidence. The land simply sinks, its watery pad pulled out from under it, and all along its surface cones of depression form. Parts of Mexico City have sunk 13 to 25 feet in elevation because of subsidence. In Phoenix, Arizona, sewer lines twist, buildings crack, freeways split. All through the American Southwest are growing fissures, some thousands of feet long and many feet deep and wide, breaking aqueducts, ruining wells. It is pointless to irrigate a field full of fissures; the water just drains away down the crack. The dropping land grows harder and more compact at the same time, harder to plow, sticky with salt when it's wet and lumpy with salt when it's dry. The fields grow more uneven, the farmer floods even deeper, with more pumped groundwater, and so on.

The cycle of irrigation, oversalting, and flooding is fueled by the crops themselves. Plants absorb the fresh water, leaving most of the salts behind, and the soil grows saltier, until it kills the crop. Irrigation is literally deadly, not only for agriculture, but for civilization: it happened in Mesopotamia, the Ganges Delta, the Bay of Bengal, in Sumeria, along the Indus, the Nile, the Jordan, Tigris, and Euphrates rivers. In the American Southwest it happened to the Hohokam, a great civilization that disappeared in the fourteenth century, leaving only hundreds of miles of irrigation canals behind. They are salty funeral marshes now, or dry plains, good only for the harvesting of salt itself. In the Old Testament salt was wholesome and salt was a killer: Abimelech razed the city of his enemies and "sowed it with salt." So we do today, to ourselves. A United Nations study estimated that half of all the farms in the world are damaged by salt.

(Don't forget where most of the harvested salt goes in this country: on the road. Never mind that it corrodes cars, ruins shoes, dissolves into city water systems and along rural highways into fields, irrigation ditches, pipelines, and always the groundwater. Salt lowers the freezing point of ice, keeping it liquid, a depression to about $-6°$ F. if the concentration is nearly one-fourth salt. The method works fine—groundwater pollution aside—in most of the United States, but is a miserable failure in Alaska, where typical winter temperatures drop far below the point where salt water itself freezes. Nevertheless, cheap and efficient salt is the method of choice in most places: almost half of the salt manufactured in the U.S. goes on the road each winter.)

THE IMPERIAL Valley of California has its own man-made salty error, the Salton Sea. It lies where the Salton Sink, a natural depression, had been, filling now and then in an equally natural cycle with overflow from the Colorado River. In the first part of this century the All-America Canal was being built, to divert water for irrigation to the fields of the Imperial Valley. In 1905 the canal was breached, and poured for more than two years into the alluvial basin. The Salton Sea was formed, and remains, already saltier than the ocean, a vast, flat, glaring plain of salt water.

The Salton Sea is only one aspect of the Colorado River, its recent history a case study of the mismanagement of water and salt. A modest river, as rivers go, the Colorado is 1,500 miles long and drops 12,000 feet from the Rockies to the Gulf of California. At almost every turn it is tapped for irrigation or dammed for a reservoir; at almost every bend it receives salt in return.

The canals off the Colorado—canals in a number of states, feeding millions of acres—water crops, trickle through the topsoil and sediment, picking up minerals and salts, and eventually seep back into the Colorado or one of its tributaries. Some of the runoff received by the Colorado is naturally salty to begin with; irrigation can turn it saltier than the sea. In the Grand Valley, the water returning to the Colorado after being used in irrigation is 28 times saltier than the water removed.

The Colorado spills into giant Lake Powell, and into Lakes Mead, Havasu, and Mojave; at each stop it evaporates, leaving silt and salts behind. The waters of the diminishing Colorado grow more saline with every mile.

If the river simply fell into the ocean, the problem would be less insistent. But the Colorado crosses into Mexico just below the Imperial Dam (and, conveniently, just where American irrigation canals stop) and it is into this leg that the last of the salty runoff, the drainage from the Wellton-Mohawk district, is dumped. The water that reached the Mexican farmers across the border was literally poisonous with salt for long years, destroying crops with its 1,500 parts of salt per million. Finally, in 1974, a treaty was signed between the United States and Mexico guaranteeing Mexico not only a certain amount of water, but water with a maximum upper limit of salt. (It is commonly assumed that the timing of the treaty, long delayed, coinciding with the crunch of oil reserves in the Mideast and the discovery of oil in Mexico, was not an accident.) For the last 15 years a pipeline nearly 40 miles long has carried the worst of the brine directly to the Gulf of California, and the sea.

One-third of the water of the Colorado, sooner or later, finds its way to Los Angeles, San Diego, and the Imperial Valley; the growing saltiness damages not only crops, marshes, and wildlife, but cars, plumbing, and machinery. Much of the immediate problem would be solved if the government bought and retired several thousand acres of the poorest farmland, the land that was never farmed before this irrigation was possible, the land with the worst drainage. But that's not the government solution, not at all.

The solution, instead, is a desalting plant near Yuma, Arizona, still years from completion, expected to cost *at least* $300 million, not including the estimated 40,000 kilowatts of electricity it will require. Part of the solution is a series of other projects upriver, costing upward of another $600 million, and ranging from the repair of drainpipes to new methods of irrigation. The farmers of the Imperial Valley have used the Salton Sea as an obvious dumping ground for their brine. Another set of plans, estimated at *another* $500 million, is being developed, to build pipelines to "natural" lakes like the Salton, brine wells to pull the salt from the groundwater before it reaches the surface, and the use of "recharge," the addition of fresh water to salty wells. The San Joaquin Valley, watching thousands of acres go out of production and overdrawing its own aquifer by a shocking 500 billion gallons a year, plans a drainage system worth another $5 billion. The net result, at least in Yuma, is water barely meeting the requirements of the Mexican treaty, at a cost of $300 for every acre-foot.

Colorado bans the deliberate evaporation of water, no matter how salty. It is simply too important a resource to lose. So the search continues for commercial uses of brine: to cool power stations, to hold heat in solar energy fields, even a brine river to carry plasticized capsules of coal from Colorado mines to ships in the Pacific. It would make a sweet purse from the salty sow's ear. In the meantime, the water table in the Great Plains drops, foot by foot by foot; the Ogallala is not only the largest aquifer in the world, it's the one disappearing the fastest. After the Dust Bowl of the 1930s, miles of sheltering trees were planted, to break the winds; those trees were cut to make way for circling, automatic sprinklers. The next time, there won't be an underwater ocean, splashing among the rocks, to revive the desert.

Certain obvious, if not ideal, solutions to the salt-and-water problem assert themselves. Irrigation can be accomplished another way. A constant tiny soak of water can be used, giving the plants only the bit they can soak up in the moment. No water is left to stand and evaporate, no water allowed to soak below the roots and pull up salt. It's harder, more labor-intensive, an old-fashioned method, to be sure. Perhaps the farmers could switch crops, and raise the saltbush, or its fellows.

In the desert of northern Chile grows a plant called the tamarugo, a tree that produces pods full of protein. Sheep like them. The tamarugo in turn likes salt, has been seen growing *through* a salt crust. Mangrove trees prefer salt to sweet. A citrus fruit called the pomelo doesn't mind salt water, nor does sugarcane or the date palm. A rubber plant called the guayule flourishes in both salt and drought, and guayule can make tires. Then there is the simple saltbush, one of a class of plants that grow vesicles filled with salt on their leaf surfaces.

The saltbush (*Atriplex halimus*) has been grown successfully in the Negev desert of Israel, in areas where little had grown before. It has a high protein content, is resistant to drought, stops erosion and the building of sand dunes, provides excellent fodder for grazing livestock. The saltbush is also highly tolerant, almost happily so, of salt. The seed sheath is usually coated in a powder of salt that has to be soaked off before the seed will germinate, and the seeds germinate very slowly. Whereas most plants germinate quickly and in large numbers, to take advantage of the immediate fuel, the salt-bush takes its time. Rainwater is scarce in the desert, and if all the seeds germinated at once they might all die in the next long drought. So a few grow, and the rest wait for another rain; comes rain, and a few more grow, leaving a few behind. And so on. Some seeds stay in the ground for years, covered with waxy, water-resistant coats as though to make the desert try its hardest to kill them.

THE THIRSTY have always gazed at seawater in a covetous hunger, seeking to split the wet molecules of water from their sea-ness, from the salty sprites of mineral ions that surround them. Even Julius Caeser tried desalting the ocean. There is something absurd and sweet about this old grandiosity of man, to make the ocean fresh, this technological optimism by the shore.

Under ideal conditions, the active desalination of seawater into fresh requires one ton of fuel oil for every 20 tons of water produced. Evaporation of one cubic *meter* of water requires 14 gallons of fuel oil, or 123 kilograms of white pine wood, or 81 kilograms of coal, or 625 kilowatts of electricity. It is a difficult, unsatisfactory, and extraordinarily inefficient process, and the amount of fuel needed in reality depends on the content of the water itself and a number of other conditions.

A half-dozen methods are in use, from an exchange of ions across filters to freezing—since ice freezes fresh, leaving salt behind. Distillation—boiling the fresh water off and catching it as steam—is a time-honored method, first used in the fourth century A.D. (Earlier methods were more troublesome: wicking seawater through wool, or filtering it, like an aquifer, through clay or sand.) The first large, modern, land-based desalination plant was built in Aruba, in the Netherlands Antilles, in 1930, and produced 625,000 gallons of fresh water every day. Now there are many large and small plants worldwide, some producing in the hundreds of millions of gallons. Dry populations quickly grow dependent on them. In one ironic twist, though, Israel is leaving desalination behind—too expensive, requiring too much fuel from Arab neighbors. Instead Israel is considering aquifer mining, of a large underground lake beneath its desert, the Negev.

Saudi Arabia, dry-mouthed but awash in fuel oil, desalinates 480 million gallons of ocean water every day, and dumps the excess salt back into the ocean, to dance off on its merry way, infesting the billion billion gallons that remain. Two of the largest plants in the world are in Saudi Arabia; one cost over $300 million to build.

With so much oil, so many dry mouths, who cares? Saudi Arabian wells are drilled more than 4,000 feet into the ground to get fresh water, and the country grows perhaps the world's most expensive wheat in its efforts to be self-sufficient. And the rich build fountains, to cool their walled gardens.

Well, even the moderately well-to-do have swimming pools in Phoenix, Arizona. The sun whips up the fresh water in an almost visible cloud of vapor, to waft across the hungry plains to the mountains, and make snow; snow for runoff, to filter through the sediment and pick up salt, to water the plants and salt the rivers. Solar evaporation of seawater to make fresh, on the other hand, seems too simple, too easy—and too slow. Of all the water in all the world, 0.005 percent is in surface water, 0.61 percent in deep aquifers, and 0.009 percent in lakes and rivers. In all the world, more than 97 percent of the water is in the oceans, awash with salt and minerals and metals and plankton and the salty urine of whales. Perhaps these long and deadly cycles—dust to dust—are just the natural response of a chaotic, heavily textured planet, knowing its limits and its needs. What was desert should be returned to desert, and water left in its ancient insulation of sand and gravel, bubbling along, in slow furrows.

16

LAKE

THE SALT Flats of old Lake Bonneville are marked with the
dents of giant heels, fingers of salt mud stretching miles into
little hills massed together as though badgered by a sheepdog.
Beyond the Flats rises the Oquirrh range of the Rockies, all
suddenly mossy green as if they'd been left in the shade too
long, slippery like creek rocks. And then the Oquirrhs come
to a screeching halt, and beyond them is the lake, a bronze,
unreflecting mirror. Color seems to rise off it like vapor, white
and gold shafts of light, opaque in a dark sky.

This is the place.

THE EVENING I arrived in Salt Lake City, the sky peeled away
from the peaks like a skin, translucent and thin, as though
there was no air until you reached that height, and between
the ground and the mountains was a pure and breathable
vacuum, a bluish shade of hot nothing. My motel was the Salt
Palace, sad and anonymous inheritor of a proud title. The
original Salt Palace was a building in downtown Salt Lake City
covered entirely with rock salt and lit at night in a glittering
show; it burned to the ground a long time ago. By the time
the taxi skirted the spidery fingers of the growing Lake and
arrived at the flat-roofed, undistinguished motel, the sky
changed. It filled with dirty black thunderheads, aggressive
and rough, and within minutes the sky filled with lightning

like a skeleton dance, sudden full-throated rumbles shaking the street and my spine. By morning the clouds and the rain were gone, gone as completely as a burst bubble, gone like struck sparks, gone like the water gone from my bowl of salt.

It is a cloudy, windy, temperate day, and waves a few feet high come steadily to a rocky, sharp-edged shore—tiny breakers scattered with whitecaps. The lake is like an ocean without tides, steady, bordered with great silent clouds of black brine flies that lie totally still, fine as grains of sand, scattering only far enough to avoid my feet. Tangled dead bushes line the shore, and the water line is foamy with the dead pupae of the flies. Gulls call like singing reeds, hovering close, hoping for scraps. And all around me the air is full of a kelpy salt-sea smell, thick and humid, brackish, visceral.

The water changes color—changes *hue*—from blue to green to gray to a most unexpected purple under a neon sky. A fine veil of cloud like a white curtain fades the blue and all around are cliffs and hills in pink and camel and a liquid pale green, distant pine and a light wheat yellow like the button-stitches of tapestry, granite grays and white rock columns, and, along the raw edge of an industrial yard, the strange oxide green of copper tailings leaching into the soil. The lake makes wind, in the interplay between cool and warm air rising from water and land, on a regular schedule: small breezes in the morning, gusty winds off the lake in the afternoon, strong winds toward the lake from evening till dawn. The wind moves clouds about, and the clouds change the lake—change its shade and shape and especially its character, so that one moment it is a placid turquoise and the next a lively dusky violet, and the next a crawling leaflike green.

Ten feet from shore, the lake barely lapping my ankles, I slip on sharp-edged rocks and slimy wood, the trunks of buried shrubs. But a slow walk out, 50 yards, 100 and still the water waist-high, and then my feet slide through the smoothest of muds. Two hundred yards and the water comes to my chest, and I turn on my back, and float.

Spread out, a dish, the water still salty enough to bounce my hands and feet and head above the surface, salty enough to hold me out in a protective palm for the sun to see. The waves slide beneath me. They rise up under my back and I slip up and down their slopes in a feathery bed. I am as light as driftwood, light as a glass eel sauntering toward shore. I am floating through the waves away from the beach; the waves rise up from below like curved plates of warm molten rock, like whole great spoonfuls of grassy earth rippling under me; they lift and slip away. Voluptuous obsidian, and my salt, how my salt longs for itself here.

THE VALLEY of the Great Salt Lake is flat, smooth—a recumbent plain. The salt flats to the west are said to be the only place on earth so smooth you can see the earth's curve as they pass away from you. The Rockies rise so abruptly from the valley floor that at a glance they appear to be painted backdrops, patterned in exquisite detail, hung like stage sets in artless array.

The lake itself is a shallow puddle on this alluvium; it spreads gently in all directions. The lake is a shadow of itself, the landlocked ocean called Bonneville that rose to an elevation of more than 5,000 feet some 16,000 short years ago. Bonneville was an Amerindian paradise lapping in the Rockies. The Oquirrhs were islands; the Wasatch range to the east a meandering series of fjords. Bonneville burst its seams through a mountain pass into Idaho and the Snake River, and now all around the broken ring of mountains are faint terraces, Bonneville's old shorelines hundreds of feet above me.

The Great Salt Lake is salty because of Bonneville's disappearance, like the final sip of broth in a bowl, overfull of season. It has no outlet but the plain, no rivers that drain it or waterfalls to cascade down. When the snowpack in the Rockies is small, the Great Salt Lake shrinks with summer heat and grows saltier; when the snow is heavy, it melts into the lake in the spring and the lake grows, larger and fresher at once.

Some people believe that the salty dust of the flats is sucked into clouds by the wind and seeds them as they pass east, and that when these pregnant clouds hit the sheer wall of the Wasatch they burst with the sweet powdery snow skiers live for. In 1983 and 1984 the Wasatch were buried deep in record snows—but the records are very new in the world's time scale. Are these fluke winters or a simple curve in the pattern?

The Great Salt Lake rose more than eight feet in two years, without a slope to climb, oozing in all directions, and changed: changed its depth, size, taste, and especially its salinity. For a long while the Great Salt Lake was a near twin of the Dead Sea, an almost supersaturated solution of 25 percent salt. Sodium and chloride and other minerals filled the lake in all their parts-per-million excess: a crowd of water pieces scattered around pieces of sodium and chloride like a confusion of confetti. The lake flooding changed 25 to a bare 5½ percent, still nearly twice the salt of the ocean, but a tremendous change. Most dramatically, the Great Salt Lake grew. In 1982 it covered about 900 square miles; two years later it was 2,300 square miles broad, and bigger than the state of Delaware.

Lake water laps all around the freeway and railroad tracks on the south shore. It is like driving across a threatened dike. When the water subsides in the west the freeway is surrounded by white; the rains cause salt in the soil to rise and crystallize atop the muddy pans.

The freeway has been raised several times. The railroad trestle is continually rebuilt. The massive evaporation ponds of the salt companies are flooded; under a thin layer of silt they hold millions of tons of nearly pure sodium chloride. Almost 400,000 acres of marsh are salted silent now, and the millions of birds that used to linger there gone—though pelicans and whistling swans have been seen. The ground is itself saturated now, and all the new melting snow runs off as though off a duck's oily back, into the lake, and the lake rises, and slides toward the city. If it rises many more feet, it will flood the runways of Salt Lake's international airport, and no one will be able to do anything about it.

SALT LAKE has been the same size, on a geologic scale, for 8,000 years. What seem to be sudden and inexplicable changes are tiny movements—wee fingers of water, droplets scurrying across the high desert floor. In this brief era of the lake, it has twice risen to what is now downtown Salt Lake City, and was not far from it when the Mormons first settled the valley in 1847. But human memory is short; by the time the lake reached its modern low in the mid-1960s, salt beds, a dozen resorts, water treatment tanks, a suburb, the freeway, two railroad trestles, and a major island causeway had been built below previous recorded elevations, with the pure optimism of the person who believes nature is on the side of human progress. Some claimed that the lake was simply drying up, draining away somewhere, without evidence to support such a notion. Plenty of people would be relieved, it seems, and only a few sorry.

What becomes a lake full of salt? A few brine shrimp, a million million brine flies, and a resort or three. The Great Salt Lake has islands (they come and go, too, with the changing waters, turning from island to peninsula and back in a few months). One island is a haven for gulls, another—long used to range cattle, until the causeway was washed out—houses a herd of buffalo. As a natural wonder,the lake has had little appeal for many locals, and has generally meant three things: a source of salt and other minerals, an attraction for tourists, and trouble.

Still, the lake offered something unique. The water at its peak of salinity was so heavy and dense that it was waveless except in hard winds, and then the waves had unexpected power. Local yachting and sailing fans designed boats with special keels to help the boats move through the dense waters—thick and yet full of buoyancy, so a boat bobbed high and off-kilter, but could make little headway. A boat held at anchor grew more inert with each day; the anchors grew crystals of salt, and then the crystals grew crystals until the anchor itself was embedded in rime.

Swimmers came to "take the waters," and waded through the thick shallows—diving into the lake was like diving into rock—and then turned on their backs, and up floated their legs to the thigh, up floated their arms to the shoulder, up floated their heads. A swimmer couldn't sink—could hardly paddle—but it was easy enough to drown, choking on an unexpected and suffocating lungful of brine.

Out of the lake came salt, salt by the ton, the thousands and millions of tons—and alum, magnesium, potash, and more. The lake was ringed with thousands of acres of squared-off rainbow ponds, this one red, that green with different algae, speckled with the dainty, high-dipping steps of sandpipers, feeding on tiny shrimp.

Men still working salt on the lake talk of the old days, and the old men of salt, the idyllic days of memory when salt was shoveled by teenagers like snow. "The far-stretching beaches shine like long strips of damask," wrote one author in the 1940s. "Sand and salt, sand and salt—dazzling white as far as the eye can reach. . . . Day after day, twenty-four hours a day, the epic process goes on—continuous, rhythmic, systematic." This wild thing of a salt lake has never failed to surprise. Winter storms blow a briny spray through the air, and it settles on electrical poles, solidifying to a frost, and the salt conducts: great arcs of spitting, angry flame leap out, pole to pole, and burn them down. It makes devil winds, sudden uproars of 75 and 100 miles an hour; it makes fog that hugs low and stays long; people have witnessed balls of fire, icebergs, storms of salt, and tiny hurricanes, and, say a few, a salt-loving monster like a giant alligator. But its strangest hours may be gone, diluted into normality.

The flooding added water, but not salt. The lake has freshened, diminished, become a bit more ordinary. The brine shrimp are having a hard time of it, and the lake even boasts a few brave small fish. The brine flies remain, but most of the marinas are gone, battered by lighter but much bigger waves, and so many of the boats are gone. The causeways are gone, several businesses are gone, much of the evaporation acreage is under water (and under silt, and under the silt is a treasure of salt). The freeway's been gone more than once, and the railroad, and the marshes and the birds are gone. But worst of all to me, an interloper seeking season, is that its absolute saltiness is gone. I want to line up truckloads of the stuff, dump them in, take a shovel to the dirty, undignified hillocks of salt by the factories, and heave the salt waterward; dissolve it, melt all this solidified fluid, break the dikes, make saline.

Everyone I speak with dismisses the lake with a kind of annoyed boredom. An eyesore, trouble, a rancid backyard embarrassment, the lake. "Go see the mine," I'm told. "Now the mine's *really* something." "I've never met anyone who saw the mine," says a tour guide, "who failed to be impressed."

And so I go, in a spanking clean tour van with a chirpy guide named Todd and a couple from Louisiana, where everything is bigger, smaller, hotter, colder, wetter, or dryer, than here. We drive away from the lake, south, swinging around a low flat butte of copper tailings dribbling pale jade green, through golden wheat fields toward the bosky Oquirrhs, and the Kennecott copper mine, the world's first open-pit copper mine, and, as Todd cheerfully adds, "Maybe the biggest man-made hole on Earth."

We wind up and up, past company houses in the company town, through a gate, past a guardhouse and up and up, and the higher we go, the larger things are: dump trucks with tires as big as my garage, the truckbeds deeper than my height, the cabs square and formidable and authoritative. We pass a train heading toward a tunnel, each boxcar heaped with two dozen tons of ore, boxcar after boxcar, and the mouth of the tunnel a great dark gash above them. And then we come round a curve on the edge of the mine, and stop.

This was a mountain, I know, once as heavy and huge as the verdant peaks that surround it high above. It was a mountain, and now it's a hole, scalloped in a spiral of roads, each descending to the next; hundreds of them pink and dusty, circling, circling the gradual cone of the pit. Little trains and miniature trucks crawl down the rim, tiny and vulnerable, and line up beside pencil-thin cranes; each crane loads 24 tons of rock. The curling lines like the spinning grooves of a record winding each into the other dip all the way to the bottom, to a voluptuous pool of poisonous turquoise water thick with copper, and a mirror of the brilliant emerald on the mountains. It was a peak, and now it is a split hymen, a wide-mouthed cavity being reamed ever deeper. I sit perched on the edge of an enormous wound, a gouge of raw earth tissue buried for eons, hidden, and now exposed.

A recorded voice speaks every few minutes from a loudspeaker, reciting the history of the mine and its many accomplishments, the data of cubic yards removed, tonnage taken, ore relieved of its birthplace. A dumptruck tire embedded in concrete sits loftily beside a hunk of copper-woven rock. People mill around, waiting their turns at the 25-cent telescopes, wait to focus on a crane arm slowly waving 48,000 pounds of rock across an arc. The Lions sell souvenirs here, jewelry and trinkets and bits of the mine. I buy a postcard of the lake, covered in glaring orange copper sheath, and a small bit of ore, verdant and sparkling, glued to a tiny copper shovel.

In 1893 the Saltair pavilion opened on the south beach of the Great Salt Lake, a massive Victorian resort with a trace of the Indian in its round domes, like a wooden Taj Mahal brought to the level of recreation. A 20-minute railroad trip from downtown Salt Lake City brought bathers, dancers, debutantes, and ne'er-do-wells by the thousands, to rent bathing suits and swim, ride the roller coaster, then shower and dance on the spring-mounted dance floor till midnight. Saltair was built over water with staircases leading directly into the lake, and lit at night to incandescence.

In 1925 Saltair was destroyed in a fire, and not rebuilt until 1929. The Depression, the Second World War, the changing culture, made quaint beach resorts less attractive. It closed in 1968 and burned in 1970.

And then was built again, in 1983, and flooded a few months later.

Now Saltair is sunk, its revisioned cupolas and streamers poking bravely out of the waves, its giant water slide toppled in pieces. The parking lot is a marina now (the old marina, a few miles west, is twelve feet under) and at one end a house-boat bounces gently, hawking salt crystals and salt taffy and postcards of the old Saltair—and postcards of the flooding, the waves hitting the back wall, the dance floor deep in water with the staff paddling through in canoes. Waves lap at the upper windows, the black clouds of brine flies scatter lazily at my approach and then settle to the foam again. Perhaps Saltair in all its tribulation is a kind of fata morgana, an image doubled, upturned, stacked, a mirage of human control of something as disorderly and dutiless as a giant salty mere.

I climb into the prow of a powerboat and we coast past the sunken, pummeled resort, all bruised and broken. Outside the marina dike the sunny day turns bouncy and rough, and the boat hops and falls and rocks from side to side, cutting through the water. It clambers up hillocks of water and tumbles down, like a soapbox derby car on the far meadow, riding high and flopping on its belly down. The boat hits the hard water like cement, wham! and I am battered about, soaked with spray and the bitten-off tops of waves. My arms, hands, legs, my face are coated in sandy white; my hair is thick with it, matted and slippery and all salt. I can't escape the taste, the smell, the stroke of soft salt water on the skin. The water itself seems alive, fingering our little boat, half-interested, its lake-mind deep and dark and distracted.

I HAVE ON the table before me two small things: my bit of copper ore, and a palm-sized crystal of salt. The copper shimmers, wetly catching and turning the light—the colors of a bruise, dark blue and violet, driblets of maroon and milky green. It is a gorgeous color altogether, the subtle shade of a dark iris in full blossom, furry. The pure copper base is as warm as fire, sleek, polished orange with the hue of a redhead. And the salt—well, it's soft-edged, with an uncertain border; I have to hold it close to find the outline. Turned this way and that in the sun, its different faces emerge: one side rough small bumps, imperfect, another side a geometric progression of cubes one inside the other, a white Mondrian. All white, but the white of glass, or ice—a transparent white, the white of diamonds, or white sand, a quartz white, a snow white, the white of the noonday sun in the eyes.

Not long ago I had a dream, a wonderful dream, the kind you wake from with a shivering reluctance and a hunger for more, wanting only not to be gone from it. It was a fata morgana dream, an alchemist's dream, hinting at worlds close and familiar but hidden by the thinnest and most impenetrable of veils.

I was swimming in a great bay, taking pleasure in the water's languid strokes across my skin, and in the empty vault below me in the wash. All at once I sensed the presence of a huge creature below the surface, its nature mysterious, intensely appealing. I had to see it. I swam out to the deep in determination, not quite prepared, ignorant. Where to begin? As I treaded in the water, feet hanging in darkness, I stepped on something smooth and solid and alive, like fleshy coral, and it reached for my foot and pulled me under.

I was tugged down, down, fathoms deep, so quickly the

bubbles rose from my breath in a startled column. There it was all quite light and transparent and I was without constraint or pressure. I stood at the feet of an enormous being, naked, round as a baby is round but muscled and full of power. It was a masculine creature, but not male, a childish thing that wasn't young; it was full of invitation. I was unafraid, staring up into its face far above me and its blue eyes limpid like wells of water or like a clear summer sky darkening imperceptibly into night. I was for a moment engulfed in the new sensation of being completely without fear, and then filled with joy like a slow tide welling from the ocean floor.

For a time in my sleep I was a bridge, between air and water, between conscious and occult, between solid and fluid. What were the alchemists who seemed to come so close? So close to bridging the chasm, "with their pestles and mortars, their crucibles and furnaces, the alembics and aludels, their vessels for infusion, for decoction, for cohobation, sublimation, fixation, lixiviation, filtration, coagulation, and botherations of every sort." Enviable souls, devout to the possibility of knowing, unimpressed (unlike all the rest of us) with large holes in the ground and disappearing lake shore. In a sense all this hodgepodge I've made is my own fumbling effort at the Elixir, at following those cycles through. Salt seems to me important not just because of its ubiquity, but because of its easeful change of state—as though salt *is* change, and the lack of it at once, motion quiescent, suspended flux. Salt is the nimble flurry of ions trading charges, clusters splitting, liquid erupting into shape, so that the air above the oddly transparent blue water of the Dead Sea seems to curl with anticipation: like Lazarus, to all appearances devoid of animation, but all alert, all awake, all alive.

Carl Jung, in reference to alchemy, called the heavy weight of the search on the searcher a "certain psychological condition." The ability first to see, and second to understand, the concealed webs of the world requires a particular psychic environment. It requires, as it were, that quiescent flux, the state-change of salt: a ripening, or readiness, a pregnancy of the soul. Don't we all long for a distillation of this given form back to its original? Don't we all, standing at the shore ankle-deep in foam and burnished with the salt air, wish for solvency? The certain psychological condition required, I think, is the willingness to melt into the earth's fluid the way the salt melts into water: to peel apart the layers, break our cubic borders, and dance, to saturate, turning earth and self, like salt and water, into something altogether new and different.

Something like the lake, saltwater, not salt, not water, but a new phase. A new state of existence. Bedeviled, bemoaned, the lake is utterly indifferent, vestal, cool, contained. Salt has always been lunar, female, dark—and bitter, hard, solid, earth. Salt transcends gender, integrates woman and man, and refuses entrance. Salt is animus, and anima, and the waters in which they live. If I sank in the Great Salt Lake, I wonder, would I become a salt anchor, too?

A PHYSICIST in Florida used a linear accelerator to add ions to the nucleus of a gold atom. He shot the particles at a terrific speed, whizzing toward the wee bit of preciousness hanging at the end. And the ions stuck, reformed—and made a molecule of lead.

A funny twist, that: gold makes lead. Salt makes gold, after all; salt added to an alloy of gold and silver and heated makes silver chloride, freeing the gold, which is washed with brine and left pure. Did we always know that, and forget? Does such a thing just have to be rediscovered again and again, by earnest physicists and enterprising chemists, biologists nosing out synthetic new tastes and engineers creating huge pumps to spray an outgrown lake into the desert? Perhaps, because knowledge has its phase changes too, from conscious to hidden to dreamed to lost, like vapor to solid to solution. I have a memory of knowledge, and in the very moment of knowing something I know also that I will forget, that knowledge will pass in the heat and distraction of my other desires. "Analysis shrinks," writes Hillman, comparing psychic salt to its chemistry. Dosage is important, to walk the line between preservation and decay.

Sometimes at night a thought comes, unheralded, and I consider running for pen and paper. But in the moment of imagining it I can see the writing, the hand moving so slowly across the page, making the odd agreement of scratch, the pen so languid, the language so barren and dull. The distillation lacks the flavor of the original. So I don't go, but lie under the sheets and let the newborn poetry wing away behind closed eyes, forming itself up perfectly in the dark, obedient and clear. As clear as the wet white moon not long gone from the window. As clear as the crystal and all it shines through.

When the sun comes in the cold window, of course I can't remember. That restless, fretful feeling returns, that vague regret. I go to the kitchen and pull the cover off the shallow pan my homemade salt is growing in and see the stones gleaming there, shining like the one Stone. Why call the Dead Sea dead, if this is what it holds? It shimmers with life, salt, in the delicate vibration between fluid and crystal. To call it dead is to miss half or more of the world's capricious and chemical existence, moving, moving, slow furrows from far below pushing quicker tides aside in their rise.

There are cultures who believe evildoers will be turned to salt, flash, and all at once become inert. Does the immobility punish, or the passivity? A few believe that a bad man will not be turned to salt, but *reincarnated* as salt, born into an existence in salt's form. Not a punishment, but a rehabilitation—into the evanescent, the edible, the solvent. I think of Lot's wife, and wonder: Is it such a terrible fate after all? The sun strikes the bowl, and the jagged hillocks glisten with an inner light. She is mine to taste and savor, mine to swim through, dive into, sleep under. She is mine in the blood, mine to weep, squirt, shake, ride upon, and save. She stands somewhere on Mt. Sodom, salty eyes on the slowly changing scene below, unmoved. She is a pellicle of salt and salt's innards, and I suspect she has salt's smile. Hers is a pure and crystalline life, fairy bright, ephemeral, soul's dirt, the salt of salt.

NOTES

Folklore, tales, superstitions, and cultural beliefs about salt are based on the following sources: Gertrude Jobles, *Dictionary of Mythology, Folklore, and Symbols* (New York: Scarecrow Press, 1961); Stith Thompson, ed., *Motif Index of Folk Literature* (Bloomington, Ind.: Indiana University Press, 1955); Christina Hole, ed., *Encyclopedia of Superstitions* (Chester Springs, Pa.: Dufour Editions, 1961); Wayland D. Hand, Anna Casetta, and Sondra D. Thiederman, eds., *Popular Beliefs and Superstitions* (Boston: G. K. Hall, 1981); *Funk & Wagnalls Standard Dictionary of Folklore, Mythology, and Legend* (New York: Funk & Wagnalls, 1972); and the *Encyclopedia Judaica* (New York: Macmillan, 1971).

The words of Aristotle are taken from *The Works of Aristotle,* vols. 1 and 2 (Chicago: Encyclopaedia Britannica, Inc., 1952).

PART I: UNIVERSAL

1 | OCEAN

As a foundation I relied on the following works: Cynthia Fansler Behrman, *Victorian Myths of the Sea* (Athens, Ohio: Ohio University Press, 1977); F. G. Walton Smith, *The Seas in Motion* (New York: Thomas Y. Crowell, 1973); Robert Hendrickson, *Salty Words* (New York: Hearst Marine Books, 1984); William A. Anikouchine and Richard W. Sternberg, *The World Ocean,* 2nd ed. (Englewood Cliffs, N.J.: Prentice-Hall, 1981); and Tjeerd van Andel, *Tales of an Old Ocean* (New York: W. W. Norton, 1977). An invaluable introduction to the study of sodium cycles and salinity is found in Robert Parker, *Inscrutable Earth: Explorations into the Science of Earth* (New York: Charles Scribner's Sons, 1984).

Ancient beliefs are based on G. S. Kirk and J. E. Raven, *The Presocratic Philosophers* (Cambridge: Cambridge University Press, 1957).

| NOTES |

The concept of paleoratio is described in a classic work, A. B. Macallum's "The Paleochemistry of the Body Fluids and Tissues," *Physiological Reviews*, vol. 6 (1926), pp. 316–356.

2 | FETUS

Clinical information is based on several excellent collections, including D. V. I. Fairweather and T.K.A.B. Eskes, eds., *Amniotic Fluid: Research and Clinical Application* (New York: Excerpta Medica, 1978); Samuel Natelson, Antoni Scommegna, and Morton B. Epstein, eds., *Interdisciplinary Conference on Amniotic Fluid* (New York: John Wiley & Sons, 1974); E.S.E. Hafez, ed., *The Mammalian Fetus: Comparative Biology and Methodology* (Springfield, Ill.: Charles C. Thomas, 1975); James Weiffenbach, ed., *Taste and Development: The Genesis of Sweet Preference* (Washington, D.C.: U.S. Dept. of Health, Education, and Welfare, 1977); and Morley R. Kare, Melvin J. Fregly, and Rudy A. Bernard, eds., *Biological and Behavioral Aspects of Salt Intake* (New York: Academic Press, 1980).

p. 14: "The mammalian fetus . . ." V. Hamburger, "Fetal Behavior," in Hafez, *The Mammalian Fetus*, pp. 68–81.

p. 14: "One scientist removed . . ." T. Lind, "The Biochemistry of Amniotic Fluid," in Fairweather, *Amniotic Fluid*, pp. 59–80.

p. 17: "A scientist tells . . ." Kare, Fregly, and Bernard, *Biological and Behavioral Aspects*, p. 22.

3 | METABOLISM

General background can be found in Allen I. Arieff and Ralph A. DeFronzo, *Fluid, Electrolyte and Acid-Base Disorders*, vol. 1 (New York: Churchill Livingstone, 1985); Campbell Moses, ed., *Sodium in Medicine and Health: A Monograph* (Baltimore, Md: Reese Press, 1980).

Standards for daily sodium requirements and the sodium content of food is from *Taber's Cyclopedic Medical Dictionary*, 15th ed. (Philadelphia: F. A. Davis, 1935).

p. 19: "And the eyes of Leah . . ." Rashi, note to Genesis 29:17, cited in Alexander Zusia Friedman, *Wellsprings of Torah*, vol. I (New York: Judaica Press, 1969), p. 60.

p. 23: "The experimental group . . ." Albert P. Rocchini, Kim P. Gallagher, Mark J. Botham, John H. Lemmer, Cheryl A. Szpunar, and Douglas Behrendt, "Prevention of Fatal Hemorrhagic Shock in Dog by Pretreatment with Chronic High-Salt Diet," *American Journal of Physiology*, vol. 249 (Sept. 1985), pp. H577–H584.

p. 23: "A radiology journal recently published . . ." Friedrich C. Luft, Eugene C. Klatte, Arthur E. Weyman, Richard Bloch, Laura I. Rankin, Naomi S. Fineberg, and Myron H. Weinberger, "Cardiopulmonary Effects of Volume Expansion in Man," *American Journal of Roentgenology* (Feb. 1985), pp. 289–293.

p. 24: Information on bezoars is based on work in the medical literature, specifically clinical articles by Edwin J. Zarling and Fred Goldner. General history of the use of bezoars is from F. Gonzalez-Crussi, *Three Forms of Sudden Death* (New York: Harper & Row, 1986).

p. 25: "One anthropologist, noting that . . ." Thomas W. Neumann, "A Biocultural Approach to Salt Taboos: The Case of the Southeastern United States," *Current Anthropology*, vol. 18, no. 2 (June 1977), pp. 289–308.

p. 26: "In 1940 a 3½-year-old boy . . ." Lawson Wilkins and Curt P. Richter, "A Great Craving for Salt by a Child with Cortico-Adrenal Insufficiency," *Journal of the American Medical Association*, vol. 114, no. 4 (Jan. 27, 1940), pp. 866–868.

p. 28: "One psychiatrist describes treating . . ." W. V. R. Vieweg, W. T. Rowe, J. J. David, and W. W. Spradlin, "Oral Sodium Chloride in the Management of Schizophrenic Patients with Self-Induced Water Intoxication," *Journal of Clinical Psychiatry*, vol. 46, no. 1 (1985), pp. 16–19.

p. 29: "But new research suggests . . ." Joan Arehart-Treichal, "Cystic Fibrosis Linked to Chloride Ions' Inability to Cross Certain Cells," *JAMA*, vol. 252, no. 8 (Nov. 9, 1984), pp. 2519–2521.

p. 29: "One more word on chloride . . ." Centers for Disease Control, *Morbidity and Mortality Weekly Report,* Aug. 3, 1979.

4 | URINE

General information is from Robert W. Schrier, ed., *Renal and Electrolyte Disorders*, 3rd ed. (Boston: Little, Brown & Co., 1986); Joan Luckmann and Karen Sorensen, *Medical-Surgical Nursing*, 2nd ed. (Philadelphia: W. B. Saunders, 1980); A. M. Harvey, R. J. Johns, V. A. McKusick, A. H. Owens, and R. S. Ross, eds., *The Principles and Practice of Medicine*, 21st ed. (Norwalk, Conn.: Appleton-Century-Crofts, 1984).

5 | BLOOD PRESSURE

I used several major sources, including Mohinder Sambhi, ed., *Fundamental Fault in Hypertension* (Boston: Martinus Nyhoff, 1984); Richard D. Mattes, "Salt Taste and Hypertension: A Critical Review of the Literature," *Journal of Chronic Diseases*, vol. 37, no. 3 (1984), pp. 195–208; Morley R. Kare, Melvin J. Fregly, and Rudy A. Bernard, eds., *Biological and Behavioral Aspects of Salt Intake* (New York: Academic Press, 1980); and an extensive survey of the literature, especially the medical journals *Hypertension, Clinical and Experimental Hypertension, Circulation,* and *Journal of Chronic Diseases*. I also found useful Campbell Moses, ed., *Sodium in Medicine and Health: A Monograph* (Baltimore: Reese Press, 1980), and the lucidity of Mordecai Blaustein's numerous articles.

p. 41: "by 1925 a physician was refuting . . ." J. A. MacWilliam, "Blood Pressures in Man, Normal and Pathological," *Physiological Reviews*, vol. 5, no. 3 (July 1925), pp. 303–335.

p. 41: Titles of academic articles are all from *Journal of Cardiovascular Pharmacology*, vol. 6 (1984).

p. 41: "We are confronted with a bewildering..." Mordecai P. Blaustein, "Sodium Transport and Hypertension: Where Are We Going?" (editorial), *Hypertension*, vol. 6, no. 4 (July–Aug. 1984), pp. 445–453.

p. 42: "The Yanomamo Indians live..." William J. Oliver, Edwin L. Cohen, and James V. Neel, "Blood Pressure, Sodium Intake, and Sodium Related Hormones in the Yanomamo Indians, a 'No-Salt' Culture," *Circulation*, vol. 52 (July 1975), pp. 146–151.

p. 43: "The Papuan people of New Guinea..." P. F. Sinnett and H. M. Whyte, "Epidemiological Studies in a Total Highland Population, Tukisenta, New Guinea," *Journal of Chronic Diseases*, vol. 26 (1973), pp. 265–290.

p. 44: "a group of Nepalese villagers..." L. Berghmanns, F. Kittel, and M. Kornitzer, "Blood Pressure Distribution and Electrolyte Excretion in a Sample of the Nepalese Population," *Acta Cardiologica*, vol. 39, no. 1 (1984), pp. 1–7.

p. 44: "such comparisons work only between populations..." Hugo Kesteloot, "Epidemiological Studies on the Relationship Between Sodium, Potassium, Calcium, and Magnesium and Arterial Blood Pressure," *Journal of Cardiovascular Pharmacology*, vol. 6 (suppl. 1, 1984), pp. S192–196.

p. 45: "continuous spectrum... which is not only inherited..." F. Skrabal, L. Hamberger, and M. Ledochowski, "Inherited Salt Sensitivity in Normotensive Humans as a Cause of Essential Hypertension: A New Concept," *Journal of Cardiovascular Pharmacology*, vol. 6 (suppl. 1, 1984), pp. S215–223.

p. 47: "A three-year Scandinavian study..." P. Pietinen, A. Tanskanen, A. Nisinen, J. Tuomilehto, and P. Puska, "Changes in Dietary Habits and Knowledge Concerning Salt During a Community-Based Prevention Programme for Hypertension," *Annals of Clinical Research*, vol. 16 (suppl. 43, 1984), pp. 150–155.

p. 48: "The author of one report from 1950..." John H. Talbott, "Use of Lithium Salts as a Substitute for Sodium Chloride," *Archives of Internal Medicine*, vol. 85, no. 1 (Jan. 1950), pp. 1–10.

p. 48: "The fate of ingested glycinamide..." M. Sternberg, D. A. Cornelius, N. J. Eberts, F. J. Schwende, and J. P. C. Chiang, "Glycinamide Hydrochloride, a Compound with Common Salt Flavor," in Kare et al., *Biological and Behavioral Aspects*, pp. 31–32.

p. 49: "Two were not only as salty..." "Salt-Free Salt," *Nutrition Reviews*, vol. 29, no. 2 (Feb. 1971), pp. 27–30.

p. 49: "Even the gourmet's mentor..." M. F. K. Fisher, *The Art of Eating* (New York: World, 1954), p. 698.

6 | SWEAT

A number of books provide information regarding thirst, heat acclimation, and the sweat gland. These include the ever-useful Campbell Moses, ed., *Sodium in Medicine and Health: A Monograph* (Baltimore: Reese Press, 1980); Robert D. Shaw, ed., *Environmental Physiology* (Baltimore: Butterworth and Co., 1974); G. Edgar Folk, *Introduction to Environmental Physiology: Environmental Extremes and Mammalian Survival* (Philadelphia: Lea and Febiger, 1966); Russell D. Butcher, *The Desert* (New York: Viking, 1972); George Uwe, *In the Deserts of This Earth* (New York: Harcourt Brace Jovanovich, 1976); and Lucy Kavaler, *A Matter of Degree* (New York: Harper & Row, 1981). I also lean on Richard Dobson's succinct "The Human Eccrine Sweat Gland," in *Archives of Environmental Health*, vol. 11 (Oct. 1965), pp. 423–429. Information on birds is partly based on Lars Löfgren, *Ocean Birds* (New York: Alfred A. Knopf, 1984).

p. 52: "Thirst is the inner consciousness . . ." Jean-Anthelme Brillat-Savarin, *The Physiology of Taste*, trans. by M. F. K. Fisher (New York: Harcourt Brace Jovanovich, 1949).

p. 55: "Writes one researcher . . ." E. R. Buskirk, "Sodium and Adaptation to High Environmental Temperatures, Work, and Athletics," in Moses, ed., *Sodium in Medicine and Health*, pp. 75–90.

p. 56: Information on sweat lodges is based partly on Selma Neils, *The Klickitat Indians* (Portland, Oreg.: Binford & Mort, 1985), and Gladys and Reginald Laubin, *The Indian Tipi*, 2nd ed. (University of Oklahoma, 1977). The quote "The noise of the steam . . ." is from Laubin.

p. 57: "Elaine Morgan hazards . . ." Elaine Morgan, "Sweaty Old Man and the Sea," *New Scientist*, March 21, 1985, pp. 27–28.

p. 61: "Claude Paque, a French anthropologist . . ." Claude Paque, "Saharan Bedouins and the Salt Water of the Sahara: A Model for Salt Intake," in Morley R. Kare, Melvin J. Fregly, and Rudy A. Bernard, eds., *Biological and Behavioral Aspects of Salt Intake* (New York: Academic Press, 1980).

p. 61: "He describes the first celebration . . ." Daniel Hillel, *Negev: Land, Water, and Life in a Desert Environment* (New York: Praeger, 1982).

7 | GEOLOGY

I owe a particular debt to Christopher Talbot and Martin Jackson's wonderful article, "Salt Tectonics," *Scientific American*, Aug. 1987, pp. 70–79, which describes salt dome extrusion and laboratory modeling in depth. I also used Robert Parker's lucid *Inscrutable Earth* (New York: Charles Scribner's Sons, 1984) and his later work, *The Tenth Muse* (New York: Charles Scribner's Sons, 1986), and James Gardner's *Physical Geography* (New York: Harper & Row, 1977). A portion of the mining information is based on literature of the Morton Thiokol Corporation, and Cedric E. Gregory, *A Concise History of Mining*

(New York: Pergamon Press, 1980). I would highly recommend the academic history, Robert P. Multhauf's *Neptune's Gift: A History of Common Salt* (Baltimore: Johns Hopkins University Press, 1978), for a detailed study of production history.

p. 64: "all the serried ages . . ." A. B. Macallum, "The Paleochemistry of the Body Fluids and Tissues," *Physiological Reviews*, vol. 6 (1926), pp. 326–356.

p. 65: "The dome beneath . . ." Multhauf, *Neptune's Gift*, p. 205.

p. 69: "More ominously, a report by two geologists . . ." *Science News*, Nov. 8, 1983.

8 | SPRINGS

American salt production history is partly based on Garnett Laidlaw Eskew's old but finely detailed *Salt: The Fifth Element* (Chicago: J. G. Ferguson & Assoc., 1948), now distributed by Morton Thiokol. The more stringent work by Robert P. Multhauf, *Neptune's Gift: A History of Common Salt* (Baltimore: Johns Hopkins University Press, 1978), describes and compares international methods of production throughout history.

p. 73: "The bubbling water . . ." Multhauf, *Neptune's Gift*, p. 126.

PART II: APPETITE

9 | TASTE

Here again I was glad of Morley R. Kare, Melvin J. Fregly, and Rudy A. Bernard, eds., *Biological and Behavioral Aspects of Salt Intake* (New York: Academic Press, 1980). Harold McGee's delightful book, *On Food and Cooking: The Science and Lore of the Kitchen* (New York: Charles Scribner's Sons, 1984), was useful both here and in the following chapter. Donald W. Pfaff, ed., *Taste, Olfaction, and the Central Nervous System* (New York: Rockefeller University Press, 1985), was helpful, as was Y. A. Vinnikov, *Sensory Reception: Cytology, Molecular Mechanisms and Evolution* (New York: Springer-Verlag, 1974). I also use information from Carl Pfaffman, ed., *Olfaction and Taste: Proceedings of the Third International Symposium* (New York: Rockefeller University Press, 1969), and James Weiffenbach, *Taste and Development: The Genesis of Sweet Preference* (Washington, D.C.: U.S. Dept. of Health, Education, and Welfare, 1977). Lastly, I am indebted for general background to G. E. W. Wolstenholme and Julie Knight, eds., *Taste and Smell in Vertebrates* (London: J. & A. Churchill, 1970).

p. 84: "like shingles on a roof . . ." Lloyd Beidler, "A Theory of Taste Stimulation," *Journal of General Physiology*, vol. 38 (1955), pp. 133–139.

p. 86: "One study compared a group of people . . ." Gary K. Beauchamp, "The Human Preference for Excess Salt," *American Scientist*, Jan.–Feb. 1987, pp. 27–33.

p. 87: ". . . the surface pressure of charged phospholipid . . ." Joseph Brand and Douglas Bayley, "Peripheral Mechanisms in Salty Taste Reception," in Kare et al., *Biological and Behavioral Aspects.*

p. 88: "the body's portable version . . ." McGee, *On Food and Cooking.*

p. 91: "In 1963 a hospital technician . . ." Laurence Finberg, "Mass Accidental Salt Poisoning in Infancy," *JAMA*, vol. 184, no. 3 (April 20, 1963), pp. 187–190.

p. 92: "The taste researcher Gary Beauchamp . . ." Beauchamp, "The Human Preference."

p. 92: "Writes one discomfited professor . . ." Discussion in Lewis P. Lipsett, "Taste in Human Neonates: Its Effect on Sucking and Heart Rate," in Weiffenbach, *Taste and Development*, pp. 125–142.

p. 93: "In 1956 the scientist Hans Kaunitz . . ." Hans Kaunitz, "Causes and Consequences of Salt Consumption," *Nature*, Nov. 24, 1956, p. 1141.

p. 100: "He would spread the bread . . ." M. F. K. Fisher, *The Art of Eating* (New York: World, 1954).

p. 101: "It has been described in pathetic inadequacy as . . ." Michael O'Mahony and Ishii Rie, "Do You Have an Umani Tooth?" *Nutrition Today*, May–June 1985, p. 4.

p. 102: "In comparing the experience of taste with seeing . . ." Robert Erickson, "Definitions: A Matter of Taste," in Pfaff, *Taste, Olfaction.* Emphasis his.

p. 104: "One researcher has found 13 . . ." Susan C. Schiffman, "Contribution of the Anion to the Taste Quality of Sodium Salts," in Kare et al., *Biological and Behavioral Aspects*, pp. 99–110.

p. 105: "One experiment just attempted to find . . ." Marianne Gillette, "Flavor Effects of Sodium Chloride," *Food Technology*, vol. 39, no. 6 (June 1985), pp. 47–53.

p. 105: "In 1925 a physician named Henry Green wrote . . ." Henry Green, "Perverted Appetites," *Physiological Reviews*, vol. 5 (July 1925), pp. 336–348.

p. 106: "There is a single case on record . . ." Michael D. Shapiro and Stuart L. Linas, "Sodium Chloride Pica Secondary to Iron-Deficiency Anemia," *American Journal of Kidney Diseases*, vol. 5, no. 1 (Jan. 1985), pp. 67–68.

10 | HUNGER

A number of useful histories of cuisine are available, and provide background for this chapter: Bridget Ann Henisch, *Fast and Feast: Food in Medieval Society* (University Park, Pa.: Pennsylvania State University Press, 1976); Jean-François Revel, *Culture and Cuisine: A Journey Through the History of Food* (New York: Doubleday, 1982); and Reay Tannahill, *Food in History* (New York: Stein & Day, 1973), are among them. I also am in debt to Frieda

Arkin's *Kitchen Wisdom* (New York: Holt, Rinehart and Winston, 1968) and *More Kitchen Wisdom* (New York: Holt, Rinehart and Winston, 1982). Fernand Braudel's remarkable work *The Structures of Everyday Life* (New York: Harper & Row, 1979), volume I of *Civilization and Capitalism: 15th–18th Century*, provided many telling details, as did Barbara Ketcham Wheaton's *The French Kitchen and Table* (Philadelphia: University of Pennsylvania Press, 1983).

Some historical details are found in Garnett Laidlaw Eskew, *Salt: The Fifth Element* (Chicago: J. G. Ferguson & Assoc., 1948), and in Tjeerd van Andel, *Tales of an Old Ocean* (New York: W. W. Norton, 1977). Lastly, I owe a debt to Harold McGee's *On Food and Cooking: The Science and Lore of the Kitchen* (New York: Charles Scribner's Sons, 1984) for information in this section as well.

pp. 113–114: This version of Gandhi's salt march is based on Homer A. Jack's *The Gandhi Reader* (Bloomington, Ind.: Indiana University Press, 1956) and Ved Mehta's *Mahatma Gandhi and His Apostles* (New York: Viking, 1976).

p. 114: "The salt itself, wrote Nehru later . . ." in Tannahill, *Food in History.*

p. 121: "In macrobiotics salt balances . . ." Michio Kushi with Marc Van Cauwenberghe, *Macrobiotic Home Remedies* (New York: Japan Publications, 1985).

p. 121: "He wrote to his wife . . ." Evan Jones, ed., *A Food Lover's Companion* (New York: Harper & Row, 1979).

p. 123: "chandeliers hung from the roof . . ." Rafael Pumpelly, *My Reminiscences*, vol. I (New York: Henry Holt & Co., 1918).

p. 123: "One room is a memorial . . ." Gordon Young, "Salt: The Essence of Life," *National Geographic*, vol. 152, no. 3 (Sept. 1977), pp. 381–401.

pp. 124–126: Information on goiters and the history of iodine in salt is from the following: Morton Thiokol Corp. literature; H. Bruce Gillie, "Endemic Goiter," *Scientific American*, vol. 224, no. 6 (June 1971), pp. 93–101; Roy D. McClure, "Goiter Prophylaxis with Iodized Salt," *Science*, vol. 82, no. 2129 (Oct. 18, 1935), p. 370; and O. P. Kimball, "The Efficiency and Safety of the Prevention of Goiter," *JAMA*, vol. 91, no. 7 (Aug. 18, 1928).

11 | CREATURES

Information in this section is based on numerous articles in the literature, particularly in the journal *Behavioral Neuroscience*, and the literature on hypertension and salt preferences cited above. Other details come from the following works: J. Green, *The Biology of Estuarine Animals* (Seattle: University of Washington Press, 1968); Ray Carleton and Elgin Ciampi, *The Underwater Guide to Marine Life* (New York: A. S. Barnes, 1956); Franklin C. Daiber, *Animals of the Tidal Marsh* (New York: Van Nostrand Reinhold, 1982); Lars Löfgren, *Ocean Birds* (New York: Alfred A. Knopf, 1984); Victor B. Scheffer, *A Natural History of Marine Mammals* (New York: Charles

Scribner's Sons, 1976); M. Peaker and J. L. Linzell, *Salt Glands in Birds and Reptiles* (Cambridge: Cambridge University Press, 1975); D. A. Denton, *The Hunger for Salt* (Berlin: Springer-Verlag, 1982); F. John Vernberg and Winona B. Vernberg, *The Animal and the Environment* (New York: Holt, Rinehardt, *Alchemy: Science of the Cosmos, Science of the Soul* (Baltimore: Penguin, 1971).

PART III: MAGIC

12 | RELIGION

Biblical passages are from the Revised Standard Version Bible and the Oxford Annotation of the Bible.

p. 137: "The Amerindians had a curious relationship with salt . . ." A. L. Kroeber, "Cultural Element Distributions XV, Salt, Dogs, Tobacco," *University of California Anthropological Records*, vol. 6 (1942), pp. 1–20.

p. 138: "An anthropologist named Neumann . . ." Thomas W. Neumann, "A Biocultural Approach to Salt Taboos: The Case of the Southeastern United States," *Current Anthropology*, vol. 18, no. 2 (June 1977), pp. 289–308.

p. 142: "The thirteenth Dalai Lama . . ." John Avedon, *In Exile from the Land of Snows* (New York: Alfred A. Knopf, 1984).

13 | ALCHEMY

A number of scholars have studied and discussed the works of Paracelsus and other alchemists, and the synthesis done by Carl Jung. Among the best works, for all of which I have been grateful in the course of this work, are: Walter Pagel, *Paracelsianism in Storm and Stress* (New York: Karger, 1984); Henry M. Pachter, *Paracelsus: Magic into Science* (New York: Henry Schuman, 1951); Arthur Edward Waite, ed., *The Hermetic and Alchemical Writings of Paracelsus*, vol. 1 (Berkeley, Calif.: Shambhala, 1976); C. A. Burland, *The Arts of the Alchemists* (New York: Macmillan Co., 1967); and Titus Burckhardt, *Alchemy: Science of the Cosmos, Science of the Soul* (Baltimore: Penquin, 1971).

Quotes and paraphrases of Carl Jung are from three books: *Alchemical Studies* (Princeton, N.J.: Princeton University Press, 1967); *Psychology and Alchemy* (London: Routledge & Kegan Paul, 1953), and *Mysterium Coniunctionis*, Bollingen Series XX (New York: Pantheon, 1963).

p. 145: "The historian Hugh Kearney describes . . ." Hugh Kearney, *Science and Change 1500–1700* (New York: World University Library, McGraw-Hill, 1971).

p. 153: "Sodium is 'a degenerated metal . . .' " Primo Levi, *The Periodic Table* (New York: Schocken, 1984).

14 | CURE

For information about homeopathy and the connections between disciplines, I am indebted to: George Vithoulkas, *Homeopathy: Medicine of the New Man* (New York: Arco Publishing, 1981); Richard Grossinger, *Planet Medicine* (Garden City, N.Y.: Anchor Press, 1980); Harris L. Coulter, *Homeopathic Medicine* (St. Louis: Formur, 1972); James Tyler Kent, *Lectures on Homeopathic Philosophy (1900)* (Berkeley, Calif.: North Atlantic Books, 1979); Edward C. Whitmont, *Psyche and Substance: Essays on Homeopathy in the Light of Jungian Psychology* (Berkeley, Calif.: North Atlantic Books, 1980); to the standard reference work of homeopathy, William Boericke's *Materia Medica*, 9th ed. (Philadelphia: Boericke & Runyon, 1927); and to the revealing personality portraits in Catherine R. Coulter's work, *Portraits of Homeopathic Medicines: Psychophysical Analyses of Selected Constitutional Types* (Berkeley, Calif.: North Atlantic Books, 1986). The chapter by James Hillman, "Salt: A Chemical in Alchemical Psychology," in Joanne Stroud and Gail Thomas, eds., *Images of the Untouched* (Dallas: Spring Publications, 1982), gave me insight into Jung's work.

PART IV: CHANGE

15 | WATER

A good source for information on disappearing groundwater and its consequences is Marc Reisner, *Cadillac Desert: The American West and Its Disappearing Water* (New York: Viking, 1986). An excellent article, by Fred Pearce, "Banishing the Salt of the Earth," *New Scientist* (June 11, 1987), provided information on the problems of the Colorado; and Charles E. Little, in "The Great American Aquifer," *Wilderness* (Fall 1987), discusses the Ogallala. Land subsistence is described in Frederick Turner, "Slowly Sinking in the West," *Wilderness* (Fall 1987).

Information on desalination is based on K. Speigler, *Salt-Water Conversion*, 2nd ed., (New York: Plenum Press, 1977); Roy Popkin, *Desalination: Water for the World's Future* (New York: Frederick A. Praeger, 1968); Paul W. MacAvoy, *Large-Scale Desalting* (New York: Frederick A. Praeger, 1969); and a number of recent articles, notably in *New Scientist*. Salt-loving plants are discussed in both Daniel Hillel, *Negev: Land, Water, and Life in a Desert Environment* (New York: Praeger, 1982), and Lucy Kavaler, *A Matter of Degree* (New York: Harper & Row, 1981).

p. 179: "In one ironic twist . . ." Helen Gavaghan, "A Saline Solution to Israel's Drought," *New Scientist*, July 10, 1986.

16 | LAKE

Technical and historical data on the Great Salt Lake is from J. Wallace Gwynn, ed., *Great Salt Lake: A Scientific, Historical and Economic Overview* (Salt

Lake City: Utah Geological and Mineral Survey, June 1980), and current information from the Utah Geological and Mineral Survey. I also appreciate Peter Czerny's labor of love, *The Great Great Salt Lake* (Provo, Utah: Brigham Young University Press, 1976), and the willingness of both Mr. Gwynn and Mr. Czerny to share with me their personal fondness for and fascination with the Lake.

p. 185: "The far-stretching beaches shine like . . ." Garnett Laidlaw Eskew, *Salt: The Fifth Element* (Chicago: J. G. Ferguson & Assoc., 1948).

p. 190: "with their pestles and mortars . . ." Samuel Brown, *Alchemy and the Alchemists*, in *Lectures on the Atomic Theory*, vol. 1 (Edinburgh: Thomas Constable Co., 1858).

p. 190: "certain psychological condition . . ." Carl Jung, *Alchemical Studies*.

INDEX